Hazardous Waste Compliance

Hazardous Waste Compliance

CLIFFORD M. FLORCZAK
JAMES E. ROUGHTON

Boston   Oxford   Auckland   Johannesburg   Melbourne   New Delhi

Copyright © 2001 by Butterworth-Heinemann

A member of the Reed Elsevier Group

All rights reserved.
No part of this publication may be reproduced, stored in a retrieval system, or transmitted in any form or by any means, electronic, mechanical, photocopying, recording, or otherwise, without the prior written permission of the publisher.

Recognizing the importance of preserving what has been written, Butterworth-Heinemann prints its books on acid-free paper whenever possible.

Library of Congress Cataloging-in-Publication Data
Florczak, Clifford M., 1951–
    Hazardous waste compliance / Clifford M. Florczak, James E. Roughton.
      p. cm.
    Includes index.
    ISBN 0-7506-7436-9 (alk. paper)
    1. Hazardous substances—United States. 2. Hazardous substances—Safety measures—Government policy—United States. 3. Hazardous waste site remediation—United States—Safety measures. I. Roughton, James E. II. Title.

T55.3.H3 F585 2001
363.72′879′0973—dc21
                                                  2001025478

British Library Cataloguing-in-Publication Data
A catalogue record for this book is available from the British Library.

The publisher offers special discounts on bulk orders of this book.
For information, please contact:
Manager of Special Sales
Butterworth-Heinemann
225 Wildwood Avenue
Woburn, MA 01801-2041
Tel: 781-904-2500
Fax: 781-904-2620

For information on all Butterworth-Heinemann publications available, contact our World Wide Web home page at: http://www.bh.com

10 9 8 7 6 5 4 3 2 1

Printed in the United States of America

Table of Contents

Chapter 1   **Introduction**   1

     1.1 Safety Culture   3
     1.2 Scope and Objective   4
     1.3 Hazard-Based Approach   6
     1.4 Organization and Planning   7
     1.5 Training   7
     1.6 Hazard Characterization and Exposure Assessment   9
     1.7 Site-Specific Health and Safety Plan   10
     1.8 Decontamination   10
     1.9 Medical Surveillance Programs   11
     1.10 Emergency Preparedness and Response   11
     References   11

Chapter 2   **Compliance Issues**   12

     2.1 Application   15
     2.2 Health and Safety-Related Programs   17
     2.3 Process Safety   18
     2.4 Interpretation and Guidance   18
     2.5 Non-RCRA-Permitted TSDs   19
     2.6 Construction   19
     2.7 Laboratory Activities   20
     2.8 Work Control System   21
     2.9 Case Histories   21
     References   25

Chapter 3   **Planning Activities**   27

     3.1 Safety and Health Program Development   27
     3.2 Roles and Responsibilities   28
     3.3 Contractor Oversight and Work Control   29
     3.4 Project Team Organization   31
     3.4.1 Project Manager   32
     3.4.2 Site Manager   33
     3.4.3 Site Health and Safety Officer   33
     3.4.4 Health and Safety Manager   36

3.4.5 Subcontractors, Visitors, and Other
On-Site Personnel ... 36
3.4.6 Occupational Physician ... 37
3.5 Communication ... 37
3.6 Security Issues ... 37
3.7 Hazard Characterization and Exposure
Assessment ... 38
3.8 Work Plan ... 39
3.9 Using Lessons Learned ... 39
3.10 Client Review ... 41
References ... 41

Chapter 4   Conducting a Job Hazard Analysis ... **42**

4.1 Why Does a Job Hazard Analysis Work? ... 42
4.2 Selecting the Jobs for Analysis ... 43
4.3 Employee Participation ... 44
4.4 Conducting a JHA ... 45
4.5 Breaking Down the Job ... 46
4.6 Identifying Job Hazards ... 47
4.7 Recommending Safe Procedures and
Protection ... 48
4.8 Revising the JHA ... 49
4.9 Process Hazard Analysis ... 49
4.10 Summary ... 52
Reference ... 53

Chapter 5   Developing a Site-Specific Health and Safety Plan ... **54**

5.1 Identifying Resources ... 54
5.2 Understanding the Scope of Work ... 55
5.3 HASP Preparation ... 56
5.4 Hazard Characterization and Exposure ... 59
5.4.1 Radiological Hazards ... 59
5.4.2 Exposure Monitoring ... 60
5.5 Chemical Handling Procedures ... 62
5.5.1 Airborne Dust ... 62
5.6 Work Zones ... 63
5.6.1 Exclusion Zone ... 63
5.6.2 Contamination Reduction Zone/Corridor ... 64
5.6.3 Support Zone ... 65
5.7 Worker Comfort Areas ... 66
5.8 Lessons Learned ... 66
5.9 Training ... 68

	5.10 Determining Applicability of Other Regulations and Requirements	69
	References	71
Chapter 6	**Development of a Site-Specific Health and Safety Plan**	**72**
	6.1 Length	72
	6.2 Specific HASP Wording	73
	6.3 Elements	73
	6.3.1 Cover Sheets	74
	6.3.2 Introduction	75
	6.3.3 Site Description/Background Information	76
	6.3.4 Project Personnel and Responsibilities	77
	6.3.5 Site Control/Work Zones	81
	6.3.6 Buddy System	81
	6.3.7 Decontamination Procedures	81
	6.3.8 Training	82
	6.3.9 Medical Surveillance	83
	6.3.10 Emergency Treatment	87
	References	88
Chapter 7	**Implementing the Safety Plan**	**89**
	7.1 Orientation	89
	7.2 Follow-Up	90
	7.3 Inspection Program	90
	7.4 Job Hazard Analysis	91
	7.5 Team Make-Up	92
	7.6 Assessing PPE	94
	References	95
Chapter 8	**Training Requirements**	**96**
	8.1 Systematic Approach to Training	96
	8.2 General Training Requirements and Guidelines	97
	8.3 Supervised Field Experience	98
	8.4 Training Certification	99
	8.5 Specific Training Guidelines	100
	8.6 Instructor/Trainer Qualification	101
	8.7 Program and Course Evaluations	101
	8.8 Emergency Response Training	101
	8.9 Lessons Learned	102
	Reference	106

Chapter 9	**Personal Protective Equipment**	**107**
	9.1 General Usage of PPE	107
	9.2 Selecting PPE for Hazardous Waste Activities	108
	9.2.1 Level A	109
	9.2.2 Level B	113
	9.2.3 Level C	114
	9.2.4 Level D	117
	9.2.5 Modified Level D	119
	9.3 Upgrading or Downgrading Levels of Protection	120
	9.4 Lessons Learned Regarding Levels A and B	123
	9.4.1 More Lessons Learned	123
	9.5 PPE Specifics for Nonhazardous Waste Sites	124
	9.5.1 General Requirements	124
	9.5.2 Compliance Requirements	125
	9.5.3 Compliance Issues	126
	9.5.4 Employee Training	127
	9.5.5 Summary	128
	9.5.6 Eye and Face Protection	129
	9.6 Equipment Limitations	130
	9.7 Respiratory Protection	132
	9.7.1 Permissible Practice	138
	9.7.2 Definitions	140
	9.7.3 Respiratory Protection Program	142
	9.7.4 Selection of Respiratory and Hazard Evaluation	144
	9.7.5 Protection against Gases and Vapors on Atmospheres That Are Not IDLH	145
	9.7.6 Medical Evaluations	145
	9.7.7 Continuing Respirator Effectiveness	146
	9.8 Lessons Learned	146
	9.9 Head Protection	147
	9.10 Foot and Hand Protection	147
	9.10.1 Lessons Learned	147
	References	148
Chapter 10	**Decontamination Activities**	**149**
	10.1 Decontamination Strategy	150
	10.1.1 Time Savings in Decontamination	150
	10.2 Acceptable Decontamination Methods	152
	10.2.1 Contact Time	152
	10.2.2 Concentration	152

	10.2.3 Temperature	153
	10.2.4 Chemical Characteristics	153
	10.2.5 Decontamination by Physical Means	153
	10.3 Using Solutions, Chemicals, and Other Materials	154
	10.4 Determining Decontamination Effectiveness	156
	10.4.1 Visual Observation	156
	10.4.2 Wipe Sampling	156
	10.5 Cleaning Solution Analysis	157
	10.5.1 Permeation Testing	157
	10.6 Defining Decontamination Areas	157
	10.7 Emergency Decontamination Procedures	157
	10.8 Identification of Decontamination Hazards	158
	10.9 Protection of Decontamination Workers	159
	10.10 Disposal Methods	159
	10.11 Equipment Decontamination	160
	10.12 Sanitation	161
	10.13 Waste Minimization	162
	References	163

Chapter 11 Emergency Preparedness and Response   164

11.1 Emergency Response	165
11.2 Applicability of Superfund Amendments and Reauthorization Act	168
11.3 SARA Title III	169
11.3.1 Emergency Planning (EPCRA Sections 301–303)	169
11.3.2 Emergency Release Notification (EPCRA Section 304)	170
11.3.3 Community Right-To-Know Reporting Requirements (EPCRA Sections 311–312)	170
11.3.4 Toxic Chemical Release Inventory (EPCRA Section 313)	170
11.4 Emergency Action Plan	171
11.5 Emergency Response Plan	172
11.5.1 Emergency Response Organization	173
11.6 Emergency Equipment and Personal Protective Equipment	174
11.7 Medical Surveillance	175
11.8 Emergency Medical Treatment, Transport, and First Aid	176
References	176

APPENDIX A OSHA Site Audits	177
APPENDIX B Choosing a Contractor/Subcontractor	213
APPENDIX C Process Safety Management Guidelines for Compliance	227
APPENDIX D Site Audit Subjects	249
APPENDIX E Commonly Used Acronyms	278
Index	**281**

Chapter 1

Introduction

Workers involved in hazardous waste cleanup, handling hazardous materials or other hazardous substances, face a more serious safety and health risk than do most construction or manufacturing operations. In addition to the typical slips, trips, and falls found in other construction or manufacturing operations, employees handling hazardous waste or chemicals may encounter a variety of other hazards including fires, explosions, and health-related issues associated with exposures to toxic substances.

Although heat-related disorders can occur in a variety of work environments, heat stress and heat-related illnesses are an especially difficult situation to handle on construction sites. These heat-related disorders become more difficult when working with hazardous materials, particularly when workers are required to wear specialized personal protective equipment (PPE). Under other conditions workers may have a potential to encounter high levels of radioactive materials mixed with hazardous material (termed "mixed waste"). Although mixed waste has been found in a variety of industries, it is considered somewhat unique to Department of Energy (DOE) sites. [1]

In this book we will concentrate on governmental regulations as they relate to hazardous waste or other hazardous materials, how to comply with specific requirements, and other best management practices (BMPs). We will focus on commercial (federal/state OSHA), DOE, and the Army Corps of Engineers operations. In addition to these requirements there may be other regulatory standards that have requirements pertinent to hazardous materials.

For example, the federal Occupational Safety and Health Administration (OSHA) regulates asbestos, lead, and other hazardous substances. It would be very difficult to provide the reader with every regulatory agency that may have jurisdiction over hazardous materials. It is not our intent to provide the reader with every detail. However, the information offered in this book can aid the reader in general compliance issues and assist in planning for safety. This, in the long run, will help to improve on-site safety performance.

Although you may not realize it, OSHA regulations are not legally enforceable at DOE facilities or Army Corps of Engineer sites. Therefore, the DOE has adopted OSHA's Health and Safety Standards Hazardous Waste Operations and Emergency Response (HAZWOPER) 29 Code of Federal Regulations (CFR) 1910.120 and 29 CFR 1926.65 and developed its own version which can be found in the DOE document O 440.1, *Worker Protection Management for DOE Federal and Contractor Employees*. In addition, the Army Corps of Engineers has adopted its own requirements as found in EM 385-1-1. These requirements, in many cases, are more stringent than OSHA's hazardous waste requirements.

In addition, the DOE has issued a variety of publications that pertain to hazardous waste. We will share some of the pertinent DOE and other information with you in a variety of places throughout this book. Much of the information that the DOE has published is useful when considering work activities involving hazardous materials. Numerous other DOE orders that outline specific requirements on safety and health programs, industrial hygiene, construction safety, occupational medicine, and nuclear safety will also be cited as appropriate for comparison.

Keep in mind that although government information is referenced throughout this book, the government has had shortcomings in the administration of health and safety at government-managed facilities. One government agency task force published a report, *Hazards Ahead: Managing Cleanup Worker Health and Safety at the Nuclear Weapons Complex*. This report noted DOE's major weaknesses, which included the following:

- The failure to establish an institutional culture that honors protection of the environment, safety, and health. The authors believe that the development and maintenance of a safety culture is a key to incident prevention and enhancing safety performance.
- The need to develop effective health and safety policies and programs for cleanup. [2]

We will be discussing many of the findings from the above report throughout this book. As we review some of the DOE's findings, we will discuss the applicability of these shortcomings to other operations. We will also compare the DOE and OSHA findings and suggest various paths forward. Planning is stressed as the basic and the first step to ensure compliance and good safety performance [1].

Although there are many references on hazardous waste/materials compliance, we have chosen to concentrate our efforts on information that has been presented in public domain documents from the DOE, OSHA, National Institute for Occupational Safety and Health (NIOSH), U.S. Coast Guard (USCG), and the U.S. Environmental Protection Agency (EPA). These documents have been summarized for readability.

In particular, we will refer to *Occupational Safety and Health Guidance Manual for Hazardous Waste Sites Activities*, and the U.S.

Department of Energy Office of Environment Safety and Health Office of Environmental Management, *Handbook for Occupational Health and Safety During Hazardous Waste Activities*. The text from the public domain documents has been condensed and has been coupled with real-life examples that will help to make this book a user-friendly reference. In addition, we have included suggested readings to provide an abundance of reference material that can be used to assist the reader in the provision of a safe work environment.

1.1 SAFETY CULTURE

As mentioned in the previous section, management is willing to accept poor performance in the areas of health and safety. This can be the case not only at DOE sites but also in private industry. Even if a company is financially sound, safety performance can take a back seat when compared to matters of sales or production. For government operations, turning a profit is not an issue. However, when dealing with private industry, the company must make money and be profitable in the long run just to survive. Whether we are dealing with a governmental agency or private industry, keeping costs down and eliminating accidents should be an important part of your operating objective.

Trying to change a safety culture (whether in a government agency or private industry) is a huge undertaking. After all, the attitude that you are trying to change has been ingrained in the management structure. Being reactive and accepting a certain number of incidents has become part of the safety philosophy. Most people really believe that "accidents just happen." The authors agree that accidents do happen, but we believe that, in almost all cases, the accidents are preventable.

In the previous section a study was cited in which DOE agreed that safety culture at some of its facilities needed to improve. The DOE is not alone in its efforts to improve safety culture. Private industry is also entering a movement to improve safety culture. Safety culture is being mentioned more often, and in mixed circles. However, safety culture is rarely defined. In an effort to describe what safety culture is, let's look at some different definitions.

The dictionary defines *culture* as "The totality of socially transmitted behavior patterns, arts, beliefs, institutions, and all other products of human work and thought typical of a population or community at a given time." An alternative definition is "The act of developing the social, moral, and intellectual facilities through education" [3].

For the purposes of this book, when we refer to safety culture we are referring to the big picture of how employees perform work as it relates to safety and health. Safety culture, simply stated, is a belief and a way of handling safety-related situations that is engrained in all employees. In a well-developed safety culture, incidents are not accepted as part of the normal way of doing business. Proactive organizations with well-developed safety cultures make sure that near misses are treated as seriously as large losses so that these losses can be avoided.

Many volumes have been written on safety culture. Many of these publications go into detail as to how to grow and maintain an active safety culture. In addition, although everyone wants a safety culture within their organization, it can be a monumental task to implement the required elements of a successful culture-building program. We believe that analogies can be drawn from the DOE studies and applied directly to all sites—government and private industry alike. OSHA has spent a considerable amount of time auditing hazardous waste sites that have been managed by both private industry and government entities. We have included in Appendix A some results of those OSHA audits. Although the information is somewhat self-explanatory, the authors have analyzed OSHA's findings and discussed key issues as they relate to safety culture and safe work performance.

The DOE and private industry have learned many lessons from years of experience in site remediation. This book will refer to selected lessons learned from the DOE, the Army Corps of Engineers, private industry, and personal experience. After reading this book the reader should have a better understanding of how to interpret the hazardous waste requirements to make sure compliance is maintained at a high level for each site-specific activity. Over and above compliance, the authors encourage the development of health and safety programs to help build a sound and workable safety culture that can be utilized across all boundaries.

1.2 SCOPE AND OBJECTIVE

This book is intended to provide the reader with some useful techniques to enhance worker protection and promote efficiency, productivity, and cost-effectiveness, along with providing the necessary quality of the work being performed. This book will further attempt to outline and define the following:

- Methods to help reduce worker injury and illness
- The scope and application of HAZWOPER
- Methods on how to implement hazardous material-related requirements through enhancements of existing programs

In addition, we will detail our discussion to help

- Clarify HAZWOPER scope and applicability to activities that may not be specifically defined in the scope of the work

- Provide some methods to help promote consistency in health and safety program development for handling hazardous materials
- Encourage a high standard for health and safety in concert with optimum productivity, cost-effectiveness, and efficiency
- Share lessons learned and help provide approaches that have been implemented on hazardous waste and other sites

Anytime hazardous materials are encountered, the potential for a mishap to occur increases. Should the hazardous materials be considered waste products, compliance issues become more important. Hazardous waste operations and work activities should be evaluated to determine if the operation should comply with HAZWOPER or other regulatory guidelines.

When it is determined that a specific operation falls under the scope of HAZWOPER, a hazard-based approach to the implementation of the various elements of the standard should be developed. When HAZWOPER is implemented, OSHA stipulates, "If there is overlap or conflict with any other standard, the provision more protective of worker health and safety should apply."

By definition, hazardous waste activities that fall in the scope of HAZWOPER include the following:

- Uncontrolled hazardous waste site
- Resource Conservation and Recovery Act (RCRA) corrective action cleanup sites
- RCRA treatment, storage, and disposal (TSD) facilities
- Emergency response operations involving the release (or substantial threat of release) of hazardous wastes and substances [2]

Some sites are easy to classify due to their inclusion on the National Priorities List (NPL), state superfund, or other regulatory list. In other cases, debate can and does arise to determine if a site should be treated as hazardous. For example, some sites commonly referred to as "brown fields" have contamination levels that are considered low. Sometimes levels of contamination are so low that exposure levels to workers do not reach action levels or permissible exposure levels (PEL). Some firms have chosen to treat low-level contaminated sites as if they fell under HAZWOPER requirements. This is a somewhat conservative approach which provides a comfort factor for management and potentially responsible parties (PRP) or other entities.

In many cases, treating sites as being hazardous waste sites can help to minimize any associated health and safety risk; if more seriously contaminated areas are discovered during site remediation, or cleanup, workers will not be overexposed based on current requirements.

Sites that may or may not fall in the scope of HAZWOPER include:

- Deactivation and certain decontamination and dismantlement (D&D) activities that do not fall under CERCLA
- Surveillance and maintenance
- Non-RCRA-permitted TSDs
- Construction
- Laboratory activities
- Research and development (R&D) activities
- Satellite accumulation sites [4]

These types of sites have been the subject of debate concerning applicability of traditional hazardous waste approaches.

1.3 HAZARD-BASED APPROACH

Hazards and their degrees vary from site to site. Over the years, hazardous waste guidelines have been used when dealing with the hazards of underground storage tank removals at the corner gas station, landfills, industrial sites, and large-scale mixed chemical or radiological sites. This hazard-based approach allows the remediation firm to use a performance-based approach when it comes to protecting workers. The greater the hazard, the more extensive the engineering controls, administrative controls, or increased levels of PPE that will be necessary. Remedial actions and associated activities at hazardous waste sites can range from low-risk, short-term to high-risk, full-scale, and long-term remediation activities [4].

Deactivation and D&D actions can range from stabilization of multiple hazards at a single site or facilities containing chemical or radioactive contamination, or both, to routine asbestos and lead abatement in a nonindustrial structure. Strategies include programs that meet compliance objectives, protect workers, and make certain that productivity and cost-effectiveness are maintained. The content and extent of health and safety-related programs should be proportionate to the types and degrees of hazards and risks associated with specific operations.

You should keep in mind the experience of your workforce along with their ability to grasp concepts or specific training. Workers who have been in the workforce for only a short time may take longer to learn certain concepts than a more seasoned worker. If the workforce is technically oriented and has some general education, the programs and training provided should be geared for that audience. On the other hand, if the workforce is transient or poorly educated, the programs and training sessions need to take these factors into consideration when developing training programs.

The hazard-based approach allows key operational hazardous waste activities to proceed in a safe and cost-effective manner. These activities may include:

- Implementing an effective access and hazard control strategy blending engineering controls, administrative controls, and use of PPE to support worker protection (see Table 1-1)
- Providing appropriate technologies and systems to outline worker and equipment decontamination activities to minimize contamination of clean areas
- Establishing a comprehensive medical surveillance program that can be used to monitor worker activities
- Initiating an effective emergency preparedness program that serves to minimize any impact to the worker, the public, and the environment [4]

1.4 ORGANIZATION AND PLANNING

Establishing an effective project team promotes comprehensive work planning, which can be used to avoid unsafe operations and unscheduled work stoppages or delays. The project team should be composed of line management and supervision, health and safety professionals, site worker representatives, engineers, other specific field personnel, or contractors and their subcontractors, as appropriate [4]. One group of workers often overlooked in the planning stages is the subcontractors. Efforts should be made to include all subcontractors because this is the group that will usually perform much of the work activity. Subcontractors have been used extensively for larger, or more complicated and hazardous, or even "dirtier" projects. Contractors and subcontractors play an increasingly important role in the safe operation of any business. We will discuss subcontractors and how they fit into hazardous waste projects in Chapter 3.

Information on how to choose the right contractor, and the proper planning prior to making the choice, are included in Appendix B. However, for now, keep in mind that subcontractors play a major part in many work activities. Obtaining input from these subcontractors at the planning stages is important to the success of any project. Subcontractors should be considered as full-time members of the project team.

In addition, project teams should encourage the use of health and safety principles in the day-to-day jobs and tasks of all workers which allows work to be done safely, on time, and within budget [4].

1.5 TRAINING

Training is the heart of any safety program, especially when the work involves hazardous substances and other related issues. Training is intended to enable the workers to recognize health and safety hazards, and to prevent incidents. As a result, training increases productivity and in some cases can improve worker morale [4].

Keep in mind that, in the past, training performed at some DOE sites represented more than 50 percent of the cost of HAZWOPER

TABLE 1-1 Summary of Access and Hazard Control Measures.

Control	Examples	Potential Advantages	Potential Disadvantages
Engineering Precludes worker exposure by removing or isolating the hazard	Ventilation Substitution Remote-controlled devices Process design and reengineering	Is most protective of worker health and safety Limits scope and application of health and safety standards Reduces specialized training requirements Does not require frequent professional health and safety coverage Eliminates PPE use Expedites work by reducing delays from decreased worker efficiency	May be costly Requires time to implement Permanent solution that may be impractical for hazardous waste activities
Administrative Eliminates or controls worker exposure by (1) managing access to hazards or (2) establishing safe work procedures	Site map and site preparation Site work zones Stay times Buddy system Security, barriers, and posting Communications Safe work plans and permits	Limits scope and application of health and safety standards Reduces specialized training requirements Eliminates PPE use Expedites work by reducing delays from decreased worker efficiency Standardizes and optimizes work procedures	May impose additional health and safety requirements Requires professional health and safety coverage
Personal protective equipment Controls degree of work exposure	Respiratory protection Protective clothing Head, eye, hand, and foot protection Additional protection (e.g., hearing)	Gives workers direct access to worksite and hazard Expedites quick entry and response	Increases worker exposure to hazard Reduces worker efficiency Requires professional health and safety coverage Requires specialized training certifications Generates waste

Adopted from U.S. Department of Energy *Handbook for Occupational Safety and Health*, June 1996, pp. 7–3.

implementation. On sites being managed by private industry, the amount spent on training is considerably less, but is certainly large when compared to non-HAZWOPER projects. Even though training has been demonstrated to be costly, a comprehensive, integrated health and safety training program is key to providing a cost-effective means of meeting those requirements. DOE recommends the use of a "systematic approach to training," in which the content and rigor of training are commensurate with the potential hazards, exposures, and work requirements [4]. Chapter 8 provides guidance to help the reader implement the training requirements.

1.6 HAZARD CHARACTERIZATION AND EXPOSURE ASSESSMENT

Hazard characterization and exposure assessment are the keys to determining the breadth of the health and safety program and associated cost. This assessment provides the information needed by the program manager to identify and design the appropriate planning on controlling worksite hazards. Along with controlling hazards, assessment results help to determine regulatory applicability [4].

In Chapter 4 we will discuss the regulatory framework and analytical tools to conduct these assessments, such as JHA (job hazard analysis), job safety analysis (JSA), safety analysis reports, process hazard analysis (PHA), and job, task, and hazard analysis. The reader needs to understand that OSHA's view on physical and chemical hazards is far reaching, as stated in the HAZWOPER standard. Note the following examples.

Section (a) (2) (i)
 "All requirements of Part 1910 and Part 1926 of Title 29 of the Code of Federal Regulations apply pursuant to their terms to hazardous waste and emergency response operations whether covered by this section or not. If there is a conflict or overlap, the provision more protective of employee safety and health shall apply without regard to 29 CFR 1910.5 (c) (1)."

Keep in mind that should a conflict exist in applicability in the CFR the more protective, or stringent applies. Typically, on a mid to large HAZWOPER site you will encounter a situation that is covered by more than one OSHA standard.

Section (c) (7)
 "Risk identification. Once the presence and concentrations of specific hazardous substances and health hazards have been established, the risks associated with these substances shall be identified. . . . Risks to consider include, but are not limited to:

[a] Exposures exceeding the permissible exposure limits and published exposure levels. ...."

Notice that published exposure levels are specifically mentioned. In the past, many felt that the only exposure limits that must be adhered to were permissible exposure limits, or PELs. This wording makes it clear that employers need to also consider reputable studies involving substances not found in the PELs.

Section (h) (1) (i)
"Monitoring shall be performed ... so that employees are not exposed to levels which exceed permissible exposure limits, or published exposure levels if there are no permissible exposure limits. ...."

Here again, published exposure levels are specifically mentioned when no PELs exist. Considering published exposure levels while monitoring is not often found in OSHA standards. The authors believe that utilizing all available hazard information can give you a better opportunity to adequately protect workers.

1.7 SITE-SPECIFIC HEALTH AND SAFETY PLAN

A Health and Safety Plan (HASP) is required before work begins and provides the link between the existing site health and safety program with the worksite-specific worker protection requirements. The HASP delineates health and safety hazards, controls, and requirements for individual activities. As previously stated, the authors believe that success on any worksite begins with the proper planning. Part of the planning process includes the design and implementation of a site-specific HASP prior to the inception of work activities. For this reason, in Chapters 5, 6, and 7 we will concentrate on various aspects of the HASP document from the development to the implementation stage. It is important to remember that the provisions of an approved HASP are part of the authorization basis and are enforceable as an extension of HAZWOPER [4]. Simply stated, all site personnel should be familiar with the HASP. The program manger, site manager, and others who may have approved the HASP share responsibility for its acceptance and enforcement.

1.8 DECONTAMINATION

Effective worker and equipment decontamination programs are critical to expedite worker egress, minimize the generation of hazardous materials, and minimize equipment replacement. Before site activities begin, containment control and decontamination programs for workers and equipment are documented in the HASP, communicated to site workers, and implemented in areas where there is a possibility for exposure to chemical, biological, or radiological hazards [4].

In Chapter 10 we discuss in more detail the overall decontamination strategy, including decontamination methods, and provide guidance for integrating nuclear and nonnuclear requirements into the decontamination process.

1.9 MEDICAL SURVEILLANCE PROGRAMS

Managers who conduct hazardous waste activities are required to implement systems to assess, monitor, and maintain records concerning employee health to minimize adverse health effects on the workforce. Chapter 6 will discuss HASP components that outline the medical surveillance requirements for hazardous waste activities. In addition, it will provide examples of how to document physical requirements, working conditions, required protective equipment, and special qualifications for all positions [4].

1.10 EMERGENCY PREPAREDNESS AND RESPONSE

Emergency preparedness should be established for the protection of the workforce and public before work can begin or be allowed to continue [4]. DOE focuses on a management system for emergency planning and response, whereas OSHA focuses on worker and responder safety. We will discuss some of these differences and offer some thoughts on integrating the requirements.

REFERENCES

1. *Hazards Ahead: Managing Cleanup Worker Health and Safety at the Nuclear Weapons Complex.* U.S. Congress Office of Technology Assessment. Washington, DC: U.S. Government Printing Office, 1993, pp. 3, 13.
2. *Management Perspectives on Worker Protection During DOE Hazardous Waste Activities.* U.S. Department of Energy, June 1996, p. 4.
3. Webster's II New Riverside University Dictionary. Boston: Houghton Mifflin, 1988.
4. *Handbook for Occupational Health and Safety During Hazardous Waste Activities.* Office of Environmental, Safety and Health Office of Environmental Management, 1996, pp. ES-3, ES-4, 1-1, 1-5, 1-6, 2-3, 2-7, 3-1.

Chapter 2

Compliance Issues

Integrating applicable OSHA, DOE, and Army Corps of Engineers standards and their corresponding documentation is a key in planning, organizing, and controlling hazards. Using a risk- and hazard-based approach to implementing specific requirements of various agencies can help to reduce duplication. Prior to determining which requirements apply, we should concentrate on determining the specific hazard. This can be accomplished through a hazard assessment, a JHA, or other selected techniques [1]. Once the hazards have been identified, the risk to workers and effect on property or the environment should be taken into consideration. Just because hazardous materials are present does not mean that all workers have to be treated as if they will be overexposed. When considering programs that are risk- or hazard-based, you can create a comprehensive, cost-effective program that should provide protection for workers and become an integral part of the project.

In an effort to keep a workforce interchangeable, site management may attempt to have all workers trained in selected topics to perform the services that they provide. This philosophy has certain advantages, such as:

- An educated workforce that can recognize a variety of hazards
- Flexibility due to cross-training
- Ease of administration

The following are some disadvantages:

- The organization has wasted resources in spending time, money, and effort in training workers who realize that they are unlikely to use the training.
- The workers who believe that they will not use the training can have a tendency to detract from the training program.

Besides detracting from the training program, a belief that the organization is wasting time, effort, and money can be very poor publicity for management in general. If workers believe that the organization is wasteful, an apathetic attitude about safety (and other areas) may develop. This apathetic attitude can be potentially dangerous.

For example, at one of the larger mixed waste sites all workers were required to receive confined space training. On the surface, this might seem like a very good idea. After all, how could there be a downside in having all workers gain a little knowledge about confined space? Unfortunately, a downside was discovered. As it turned out, this all-inclusive rule meant that everyone on site would be trained, including truck drivers. There were lectures and written lessons in the morning, and hands-on training, including rescue, in the afternoon. The rescue included having the worker wear a harness and lifeline in a room that was a mock confined space. Workers would then use a rescue winch to retrieve the worker from the mock confined space by pulling the worker through a cardboard tube on command.

As one truck driver participating in the training was being pulled through the tube, he became stuck. He called out to advise the workers that were operating the winch that he was "stuck." Unfortunately, the worker on the winch thought that the truck driver was "fooling around," and the truck driver ended up with a serious groin injury. This truck driver had more than ten years with this site and had never had the opportunity to use confined space training. In this case, awareness training would have been more appropriate than extensive training. This awareness training could have provided the driver with a little knowledge about confined space, while costing the organization a fraction of the resources as compared to the full program.

This type of situation can occur often. Some sites have specifications that call for universal training for all subcontractors. Some contract administrators have interpreted the word "universal" to be just that. In this situation, it would be likely that workers might get more training than they need.

Let's look at another example: At a dormant manufacturing facility, an outside contractor was hired to remove asbestos from a large steel storage tank. Although the facility was no longer in production, there were security guards stationed at the facility. This particular storage tank was outdoors. The bid specification did not require that the asbestos abatement be performed in a negative pressure enclosure. An OSHA compliance directive was referred to that indicates that outdoor removals without enclosures are acceptable in most situations.

A dilemma surfaces. It appears that the ongoing asbestos removal work is compliant, however, the security force has voiced health concerns. What course of action should be recommended?

As in most instances, there are a variety of ways to properly handle any situation. We offer some choices which we believe you may find helpful.

If health concerns have been raised, the first order of business might be to assess the validity of the health concerns. This assessment should include as much analytical information as possible. This might mean medical examinations coupled with blood tests, biological indices, chest X-rays, or other methods. It could also mean air monitoring, both personal and area monitoring, with any results explained to those potentially exposed.

We believe that the explanation of results is very important. Getting results that are below the detection limit or far below any PELs or action levels will sometimes go unreported or be given very little attention. We believe that any number, even zero, is well worth discussing with anyone voicing a health concern. Posting numbers and not discussing results that are below PELs may be a compliant practice, but we believe that getting to a personal level is a much better practice [2].

Training is another important issue. Workers should not begin work activity until they have been adequately trained. This training includes making workers aware of potential hazards they may encounter [3]. Training and information sharing should begin immediately if a health concern is raised. In a proactive culture, we believe that health concerns are discussed well before workers are potentially exposed. In the case that we are discussing, it was unclear if there was a requirement to train the security guards regarding the hazards of asbestos. In addition, even if there were a requirement:

- What type of training should they receive?
- Who should give the training?
- Who is responsible for providing the training?
- Who pays for it?

The answers to these types of questions are not always straightforward, especially when the security force is employed as an outside contractor. However, failing to give the security guards information and training regarding the hazards of asbestos, or arguing over logistics for an extended period of time, is likely not the best choice.

However, in this case, this situation was resolved when samples were taken and awareness training was given to the security force. Once these two items were completed, the security force became more valuable team members and became noticeably more involved in site matters.

If we are going to follow HAZWOPER principles, why should we determine if the operation falls under these requirements? The answer is simple. If we follow these principles it will help to make sure that a job is done safely. If the specific work falls in a "gray area," using HAZWOPER principles will help to eliminate controversy over any compliance issues.

How do you know if an operation falls under the hazardous waste standard? We need to answer this question before we get too deep into the realm of hazardous waste remedial activities. Whether the answer to the question is yes or no does not mean that a job does not need to be performed with trained workers, as discussed in the case history presented. No matter if the site is covered or not, the underlying principles are sound and should be used. We will discuss some of the underlying principles that are used in HAZWOPER when we discuss the requirement of handling hazardous substances.

In principle and in practice, being compliant (at a minimum) will help to protect site workers, the public, and the environment. More progressive or conservative organizations will not use compliance only as a benchmark, but will have internal requirements that are more stringent or protective. After all, OSHA standards are minimum requirements.

Let's use an example that reflects this philosophy: confined space atmospheric limits. Let's say that the regulation pertinent to acceptable oxygen levels has a lower limit of 19.5 percent (OSHA sets limits at 19.5 percent to 23.5 percent). An internal policy might choose the limit at no less than 20 percent. In another case, an organization might use the acceptable lower explosive limit (LEL) of 5 percent, as compared to OSHA's 10 percent.

This same organization may insist on fall protection at five feet instead of the six feet rule as outlined in the construction standard 29 CFR 1926.503, and so on. The point is simple. If you follow OSHA you have set minimum requirements for your operation. This is okay for some situations, but progressive organizations will set higher standards to make sure that all employees are protected to a greater extent. It is your decision, and a reflection of your company's safety program.

2.1 APPLICATION

How do we determine if a site activity is covered under HAZWOPER? There is no simple solution, but there are some simple guiding principles that can make the task of determining applicability easier. The questions we want to ask ourselves are:

- Does the activity pose a reasonable possibility for exposure? or
- Does the activity inherently expose workers to hazardous substances, or to health and safety hazards from a hazardous waste operation?

HAZWOPER applies only where exposure to hazardous substances or to health and safety hazards resulting from a hazardous waste operation is likely (see Figure 2-1). This can be determined by analysis of exposure monitoring data, hazard characterization, hazard analysis, or exposure assessment [1]. Some of the specific examples of work activities and situations will be covered later.

16 *Hazardous Waste Compliance*

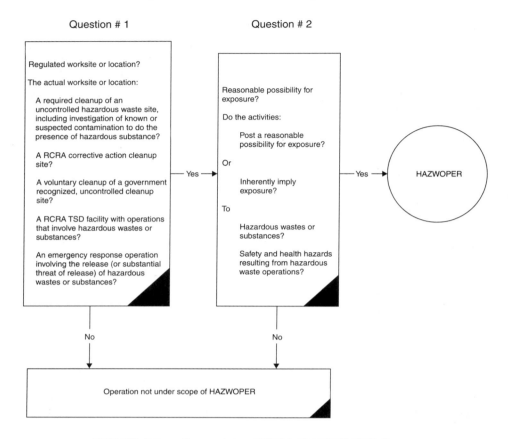

FIGURE 2-1. *Determining OSHA HAZWOPER Scope*

Making the determination of applicability of HAZWOPER is a matter of heated debate and many times becomes a legal battle. If you are lucky enough to have a management group knowledgeable in hazardous waste issues, you may consider forming a subcommittee to discuss each aspect in detail. With the team approach you usually will get a consensus of opinion. Although there appear to be fine lines of applicability and numerous gray areas, these issues have a way of working themselves out. With a management team interested and concerned about compliance, the action plan for determining applicability usually becomes obvious.

Once the decision is made that an operation is covered under HAZWOPER, the appropriate paragraphs of the standard should be applied to specific activities. Paragraphs (b) through (o) apply to environmental remediation and corrective actions, paragraph (p) applies to RCRA-regulated TSD facilities, and paragraph (q) applies to certain emergency responses to releases (or threats of releases) of hazardous wastes or substances, without regard to location [4].

2.2 HEALTH- AND SAFETY-RELATED PROGRAMS

For many DOE sites, safety, health, and environmental management is a dynamic process that typically starts with the deactivation activities (stabilizing a facility or project site). Surveillance and maintenance is an intermediate step in the process which allows required systems to operate until the facility or operation is ready for decommissioning. This leads us to the final stage, decommissioning. This stage will consist of decontamination, dismantlement, and remediation. Application of these provisions of the appropriate rule or requirement depends on the site-specific facility or operation, the associated hazards, and the potential for worker exposure to the hazards.

For large superfund sites, the process can be similar to the DOE process as described. Once the site has been adequately assessed, a remedy can be chosen. This remedy can vary but could include a removal or stabilization phase, a treatment phase, a maintenance phase, and, finally, dismantlement and decontamination phases.

For other CERCLA sites the process can be very different from the typical DOE site. The process may start with various phases of site assessments. The intermediate step may be a pilot study, followed by a pilot plant operation, or possibly a removal action or other alternative. The final steps may vary widely. However, just as in DOE sites, the appropriate rule or requirement depends on the site-specific facility or operation, the associated hazards, and the potential for worker exposure to the hazards. For the Army Corps of Engineers cleanup or oversight, the rules will most likely be even more stringent than for OSHA or DOE.

It is important to differentiate between the scope and application of a standard of practice. *Scope* determines that an operation or location is "covered" or "governed" by the standard. *Application* determines that portions (e.g., paragraphs) of the standard apply to the particular operation or location [1].

These types of analysis may exclude many routine activities from specific requirements under HAZWOPER while continuing to provide adequate and appropriate worker protection. In each case the operation should review each situation and make the best decision on how to handle the entry based on the interpretation of the particular requirements.

Certain activities conducted by DOE or the Army Corps of Engineers normally fall outside the scope of HAZWOPER. For these activities HAZWOPER concepts and principles should be used as a framework, and not as a rigid standard for their planning and conduct. The following list summarizes some considerations when determining the application of HAZWOPER as a framework for projects not strictly regulated by the standard.

- Determine if HAZWOPER needs to be applied or if applying its concepts or principles would suffice. This determination should be made by a competent individual responsible for hazardous waste activities.
- Apply all elements of HAZWOPER to environmental remediation involving radioactive wastes and materials. (Note: OSHA treats radiological and nonradiological environmental remediation activities similarly.)
- Identify jobs and tasks that require hazard analyses.
- Integrate hazard analyses to identify worker hazards and to provide a basis for specification of job and task hazard controls. (The upcoming section covering hazard characterization and exposure assessment will provide some suggestions on effective ways of conducting hazard analyses using the HAZWOPER job, task, and hazard analysis approach [1].)

2.3 PROCESS SAFETY

Another issue that sometimes comes into play is Process Safety Management (PSM). You should be aware of the issues surrounding the requirements. The process safety management practices were originally developed by leading private-sector chemical manufacturers and called "responsible care." This program refers to management practices that integrate process safety information, hazard and operability studies (HAZOPS), and other methods that may apply. In addition, health and safety plans, management of change, operating procedures, safe work practices, training, mechanical integrity of critical equipment, prestartup safety reviews, emergency response and control, investigation of incidents, and management system audits are all elements that should be considered. These systems now fall under OSHA's 29 CFR 1910.119, Process Safety Management [1]. You should refer to the intent of the standard to understand how it may apply to your particular operation. In most cases, if you are working around what OSHA refers to as Highly Hazardous Chemicals (HHC) then you will likely be covered. Check with the facility that you are working with and review the list that is detailed in the PSM standard. One of the important elements to review to understand your particular compliance status is the Total Quantity (TQ). Refer to Appendix C for more information on Process Safety Management.

2.4 INTERPRETATION AND GUIDANCE

OSHA provides guidance on interpretation, including numerous examples, in its publication *HAZWOPER Interpretive Quips* (IQs.) The IQs are policy statements abstracted from official OSHA letters of interpretation. OSHA makes it clear that decisions regarding scope should be supported by hazard characterization and exposure assessment (refer to Chapter 3). The final determination should be made by a qualified person.

When determining the scope of HAZWOPER, exposure includes two elements: the presence of a hazard and worker access to the hazard. For example, contaminated areas of a hazardous waste site potentially pose some level of health hazards.

For exposure to occur, workers should have access to the hazard (e.g., they should work in or near contaminated areas). Under normal circumstances, those workers who are prevented from entering contaminated areas (by using access controls) are not exposed to contaminated material. In many cases these workers do not fall under the requirements, provided that they are not exposed to other safety hazards as a result of the operation. Conversely, workers in contaminated areas are covered because they have access to health hazards and could be potentially exposed [1].

Safety hazards are treated in the same manner. For example, workers who work in trenches in clean areas of the site would be covered by the OSHA Excavation and Trenching Standard, Subpart P, 29 CFR 1926. Workers who work in trenches in contaminated areas would fall under both Subpart P and HAZWOPER. Workers who do not work in trenches fall under HAZWOPER only when working in contaminated areas and would not be covered by either standard when working solely in clean areas, provided they are not exposed to safety hazards resulting from hazardous waste operations.

2.5 NON-RCRA-PERMITTED TSDS

Non-RCRA-permitted TSDs and waste treatment activities not covered by RCRA (e.g., wastewater treatment facilities permitted under the Clean Water Act) are not covered by HAZWOPER, except for emergency response and some limited waste management operations. Specific HAZWOPER elements are assimilated into the existing health and safety program based on hazard analyses. Worker protection requirements are met through existing health and safety plans [1].

2.6 CONSTRUCTION

The construction industry has some unique characteristics. You may not think that you will encounter hazardous material when working on a construction project, but you must decide if there is a reasonable possibility that hazardous substances could be encountered during any intrusive activities. HAZWOPER applicability should be determined during the project's planning stage, based on hazard analyses and the possibility for exposure [1]. Construction health and safety measures stipulated should be incorporated into the HASP. Although hazardous waste applicability is usually determined during the project's planning stage, discoveries have been made during activities that were considered "construction only" and not hazardous that warranted a quick change in status. As we will discuss later, if unplanned events take place during work activity the status of a project should be revisited. The unearthing of buried drums or wastes during construction or contact with other material such as lead or asbestos has happened on too many occasions.

HAZWOPER sites are subject to the same rules and requirements as other operations. This holds true whether the site is being managed by private industry, DOE, or the Army Corps of Engineers. Identifying and implementing a project team in the early phases of the project to address health and safety issues will help to achieve seamless integration and to reduce duplication.

2.7 LABORATORY ACTIVITIES

Any site-related activities such as bench-scale laboratory and R&D activities should comply with the OSHA Laboratory Standard (29 CFR 1910.1450). R&D activities involving pilot- or full-scale field operations should comply with HAZWOPER when there is reasonable possibility for worker exposure to hazardous wastes or substances or emergency response.

There are also other conditions that should be taken into account, such as satellite, accumulation sites, non-TSD facilities, and waste management activities. Under these conditions OSHA allows conditional exemptions for small-quantity generators (i.e., those that accumulate less than 100 kilograms per calendar month) and full exemptions for storage areas housing hazardous waste for 90 days or less. With proper documentation, these conditions may not be classified as hazardous. The determination that the user makes should be based on available information. The EPA stipulates that 90-day generators require their employees to be trained to participate in emergency response activities. An emergency response plan or emergency evacuation plan is also required for each site. Emergency response provisions of paragraph (p) are applicable, depending on employee responsibilities in responding to spills [1].

If it is determined that HAZWOPER applies, a site-specific HASP should be developed. As previously mentioned, a HASP document provides the basis for a successful project. In Chapter 5 we will discuss the details for developing a site-specific HASP.

Employers should provide appropriate training and medical monitoring based on a needs analysis. Taking a common-sense approach is recommended. Certain monitoring is important to make sure that workers are physically able to perform their jobs successfully. A prudent business practice is to make sure that basic monitoring is performed for every worker. If workers may become exposed to hazardous substances, monitoring should be performed to determine their current baseline or body burden. A "fit for duty" statement, signed by an appropriate healthcare professional, should be obtained before assigning any work. Details regarding medical monitoring programs will be discussed later.

2.8 WORK CONTROL SYSTEM

Health and safety planning and implementation emphasize jobs and tasks. Many DOE or Army Corps of Engineers sites have an established work control system (WCS) that is focused at the job and task level. Workers are familiar with the WCS and understand its content because each work-task package includes checklists and permits. This is a normal part of daily work. The WCS is a practical vehicle for managing and conducting these activities and supports the HASP by providing a mechanism to accomplish the following:

- Ensure that all hazard analyses are included in the HASP
- Evaluate (proposed) tasks to verify that the safety concerns are adequately addressed
- Promote participation by workers, managers, and health and safety professionals [1]

2.9 CASE HISTORIES

Now that we have discussed some details of HAZWOPER, let's review some case histories to see how we can put this in perspective. Case histories are important because they can be used as learning tools.

Case 1: Truck Drivers Hauling Clay

If a truck driver is hauling clay fill into an exclusion zone, does this fall under the HAZWOPER standard? At some sites, it might be a requirement that ALL persons (including truck drivers) are HAZWOPER trained. However, in all likelihood, a requirement to HAZWOPER train all truck drivers would be a difficult requirement to administer. At most sites, drivers are not HAZWOPER trained. One way to avoid this training would be to require that the drivers do not drive through contaminated areas. In addition, make sure that the drivers know that they must not leave their trucks and that they should keep their windows rolled up.

Keep in mind that we are not recommending that your drivers should not be trained. On the contrary, if it is reasonable to train the drivers, it is an excellent idea. However, the truck driver population often is transient by nature. After you have invested time and money to train a driver it can be difficult to ensure that you will be able to reap the benefits from this training. Drivers can be dispatched to a variety of places for a variety of reasons. Keeping this in mind, it makes sense to manage so that the drivers would not be required to be HAZWOPER trained.

If you are confident that monitoring data indicates that these workers have no reasonable possibility for exposure to hazardous substances, this can help justify the requirement (or lack of) for HAZWOPER (and possibly respirator or other) training for drivers. Therefore, a case can be made that this type of hauling operation is not covered because the truck drivers are not exposed to hazardous materials. The truck drivers are exposed to safety hazards that are a result of the hauling operation, not the hazardous waste operations. In this case, the truck drivers must successfully complete appropriate training (e.g., the site-specific briefing, general employee training, and possibly defensive driving training), but probably not the 40-hour HAZWOPER training [1].

The procedures that truck drivers follow are documented in the HASP. A competent person should periodically monitor the hauling operation to verify that the workers continue to have no reasonable possibility for exposure. Also, keep in mind other work requirements. For example, many firms require that their truck drivers leave the cabs of their trucks and stand aside, at a safe distance, during the loading procedures. This rule is put into effect so that the materials being loaded can not injure a driver. In addition, there is excellent logic in this rule when the material is irregular. Examples of the types of materials considered irregular include scrap metal and concrete slab pieces from demolition of highway debris. The driver should definitely exit the cab while irregular materials are being loaded. What planners fail to realize is that the drivers need a safe place to stand while the truck is being loaded. This place should be close enough for the driver to observe loading, but out of the weather and far enough away to prevent injury. A competent person should occasionally ride with a driver and observe the practices, making recommendations for improvements when necessary.

Case 2: Utility Workers Servicing Electrical Equipment

When utility work is located in an exclusion zone, are workers who enter the area exposed to hazardous materials? Hazard characterization and exposure assessment performed by a competent person may show that the area surrounding the equipment and an access corridor leading to the equipment can be cleaned so that the utility workers can work in the assigned area and travel through the corridor without possible exposure to hazardous material. The work can be carried out as a normal maintenance operation.

If the area and corridor can be maintained free of safety hazards arising from the hazardous waste operation, the work probably would not fall under the requirements. In this case, the area and corridor would constitute a temporary support zone. Because the work involves electrical utilities, it would fall under the most protective standard of practice, such as OSHA's Electrical Standard or the National Electric Code (NEC). Also, there may be other requirements that apply. Administrative controls such as HAZWOPER-trained escorts are used to make certain that the utility workers are not exposed to any hazards from the operation. The procedures to be followed are documented in the site-specific HASP [1].

We must stress that you should strive to have trained electricians. The electricians, when compared to truck drivers, are not as transient a workforce. You can find many electricians who have HAZWOPER training, and you are more likely to retain the electrician should you decide to make the commitment to train the electricians. You might be surprised to find that you can locate HAZWOPER-trained electricians in the more populated areas. This workforce is more difficult to locate as you move away from larger cities. The trained worker likely feels like part of the team when management invests the resources to provide the worker extensive, appropriate safety training.

Case 3: Support Personnel

HAZWOPER does not cover clerical or support personnel, workers at the perimeter of a hazardous waste worksite, or workers engaged in construction activities in uncontaminated areas, provided they are not exposed, or have possibly been exposed, to hazards resulting from the operations. These workers would fall under the scope of other appropriate standards of practice that are more protective of health and safety [1].

Exposure or the likelihood of exposure is the key. If the likelihood of exposure of any worker (including clerical workers) exists, an assessment should be conducted. The site controls that have been designed and installed to limit access or exposure must be monitored. These controls should be installed so that there are multiple levels (dependent on the severity of the hazard). If one level fails, the next level should be sufficient to protect workers until repairs to the first level can be completed.

Again, we are not attempting to encourage shortcuts. We believe that effective, appropriate training is a key part of any project. This holds true for clerical workers also. For those clerical workers who do not get the

40-hour core training, consider a 24-hour training. If the 24-hour training cannot be performed, an extensive orientation with updates as necessary is very important. You need your clerical help to be part of the team effort. Keeping clerical help well informed can prove to be a great asset.

Case 4: Environmental Remediation Planned at an NPL-Listed Site

The worksite includes an abandoned building that has been slated for renovation for use as a storage facility for later operations. The building contains large quantities of friable asbestos in the ceiling insulation and pipe wrappings. The building also contains concrete walls covered with lead-based paint. There are no other hazardous substances or wastes present in the buildings.

For asbestos removal, the provisions of the OSHA Asbestos Standard 29 CFR 1926.1101 are more protective of worker health and safety than are the more general provisions. The HASP therefore provides that the asbestos removal tasks conducted inside the building will be performed in accordance with the OSHA Asbestos Standard. After the asbestos has been removed, the lead-based paint will be removed. Again, the provisions of the OSHA Standard for lead removal are more protective of worker health and safety than are the more general provisions of 29 CFR 1910.120. Therefore, the removal of the lead-based paint inside the building will be performed in full compliance with the OSHA Lead Standard [1].

For example, in considering workers in contaminated areas of the site who work on scaffolds, the OSHA Scaffolding Standards are more protective for safety hazards resulting from working on scaffolds. HAZWOPER is more protective for health hazards resulting from the contamination. The applicable provisions of both standards would apply to the work.

Again, we believe the more training, the better. In addition, as mentioned with electricians in a previous example, you will likely find workers who are trained and qualified to perform HAZWOPER, ACM, and lead abatement, especially if your site is near a large population center.

Also keep in mind that most asbestos abatement is closely monitored by state and local governments. Although OSHA has jurisdiction, the states and local regulators typically keep a watchful eye over ACM activities.

Case 5: RCRA and TSD Facility

An RCRA, TSD facility consists of tank farms and wastewater treatment plants handling low-level radiological wastewater. The tank farms with uncontrolled environmental releases undergo corrective actions. Do paragraphs (b) through (o) of HAZWOPER apply to the entire facility? Does paragraph (p) apply to the part of the TSD not undergoing corrective action? Would paragraphs (b) through (o) apply to routine decontamination of the TSD?

Paragraphs (b) through (o) apply only to the portions undergoing remediation. If normal operations were not affected by the uncontrolled releases, paragraph (p) would apply to those unaffected areas. Defining decontamination activities using established controls for normal operation places these activities under 29 CFR 1910.120 (p). For example, decontamination of an evaporator facility is controlled by standard operating procedures, safe work permits, and as-needed task instructions as part of the overall health and safety program.

Similarly, routine maintenance or replacement of process lines in the wastewater treatment facility would be work covered under paragraph (p). Remediation efforts to clean up leaks at the tank farms are covered under paragraphs (b) through (o) [1].

Case 6: Emergency Response Activities

OSHA clarified HAZWOPER's application to some waste management and emergency response activities. For example, drum handling and similar tasks that are controlled by operational safety procedures and that occur in a building's envelope are generally not covered. Likewise, small, localized spills (e.g., from a 5-gallon pail) that are readily controlled by workers normally assigned to the operation are generally not covered. However, large, uncontrolled spills or removals of drums that occur outside the building's envelope are covered.

This decision is one that must be made after carefully considering all of the circumstances and, of course, based on the requirements. The reportable quantity rules may come into play. The principle behind these requirements is that the more dangerous a material might be to personnel, surroundings or the environment, the smaller the reportable quantity. A competent person should review each occurrence to help determine the appropriate action. Sometimes, even a very minute spill must be reported. If doubts occur as to the applicability of the requirements, take the safe rather than sorry stance.

REFERENCES

1. *Handbook for Occupational Health and Safety During Hazardous Waste Activities.* Office of Environmental, Safety and Health Office of Environmental Management, 1996, pp. 2-1, 2-3–2-9, 5-5.

2. *Hazards Ahead: Managing Cleanup Worker Health and Safety at the Nuclear Weapons Complex.* U.S. Congress Office of Technology Assessment. Washington, DC: U.S. Government Printing Office, 1993, p. 7.
3. *Occupational Safety and Health Guidance Manual for Hazardous Waste Site Activities.* Prepared by National Institute for Occupational Safety and Health (NIOSH), Occupational Safety and Health Administration (OSHA), U.S. Coast Guard (USCG), U.S. Environmental Protection Agency (EPA), October 1985, p. 4-1.
4. 29 CFR 1910.120.

Chapter 3

Planning Activities

Some key elements that should be considered when conducting any work activities include organizational structures and project planning. Proper planning will lead to work being done both safely and efficiently [1]. Contrary to popular belief, safety and efficiency are not diametrically opposed. Safety and efficiency both have an important place in the hierarchy of project management. In a true sense, you cannot have one without the other. These elements take on an even greater role when working with hazardous materials. A project team of line management project directors, project managers, supervisors, health and safety professionals, subcontractor representatives, engineers, and worker representatives allows the structure of work to be defined and implemented in the proper manner.

We emphasize the involvement of subcontractors because, many times, numerous different subcontractors are the ones doing most of the site activities [2]. Using experienced specialty subcontractors can be the most efficient and safest way to get the job done. Useful information when attempting to choose a contractor who will perform work in a safe and healthful manner can be found in Appendix B.

For planning purposes, the importance of subcontractor participation, organization, and planning activities is important and should be stressed. Throughout the rest of this book, it is assumed that subcontractors' workers will be considered as part of the work team.

3.1 SAFETY AND HEALTH PROGRAM DEVELOPMENT

An effective health and safety program begins with management commitment to help achieve consistent worker protection. Senior management is responsible for demonstrating this commitment at all levels and encouraging workers to accept safety as an integral part of their jobs [2].

These goals cannot be realized without accomplishing the following:

- Establishing overall and specific organizational roles and responsibilities of different functions and disciplines by defining individual roles, responsibilities, accountabilities, and interfaces in the project team with matrix personnel and organizations, and between contractors and subcontractors.
- Orienting the health and safety organization toward teamwork.
- Finding solutions while avoiding confrontation.
- Demonstrating management's commitment to a safe work environment.
- Providing health and safety planning for site-specific projects, at the job and task levels.
- Bringing workers from different technical disciplines into project teams. (This will encourage employee participation.)
- Verifying that project teams have adequate technical resources (and knowledge) to complete the project or task in a safe manner.
- Incorporating lessons learned into work practices.
- Allowing completion of work safely and cost-effectively.
- Coordinating with the local emergency response team.

Relying on teamwork to integrate health and safety and line management functions for the planning and accomplishment of work activities is vital to providing a safe working environment. Health and safety excellence should be the primary mission objective [3].

3.2 ROLES AND RESPONSIBILITIES

Site-specific health and safety requirements and site personnel, including contractors, are typically held responsible for managing and conducting all activities safely. Every worker should understand that he or she is responsible for sharing in the commitment to a safe workplace. In addition, employees should perform their work in accordance with any applicable laws, regulations, contract provisions, and established site-specific requirements.

Given that multiple contractor and subcontractor organizations could be involved in work activities, senior management should address any misunderstandings concerning specific operational responsibilities and accountabilities that could cause problems in the administration of site-specific programs.

Defining responsibilities and levels of authorities should be specified in the contractual agreement. This fundamental strategy is essential for success. The more complicated the task, the more in depth the contractual agreements and site-specific plans will need to be.

Health and safety issues and worker protection should be integrated into project specifications, bid packages, contracts, and other appropriate project documentation and submittals. To provide a clear understanding of what is expected, it is encouraged that pre-bid and post-bid meetings be conducted. Health and safety professionals should be included during the planning discussions and client meetings to make sure that they understand the scope of work. Workplace reviews should be periodically performed by project management and health and safety professionals to verify the adequacy of hazard controls. These assessments should be conducted with first-line supervisors and workers, focusing on reinforcing management activities to achieve safe work practices [3].

3.3 CONTRACTOR OVERSIGHT AND WORK CONTROL

Successful project control includes understanding and anticipating organizational issues that may occur with contractors and subcontractors. Once the contractor relationship is formalized, it is then communicated to all affected personnel on the site.

Contractors and subcontractors are typically required by contract to be responsible for their own workers and should provide a level of oversight to meet all specifications. The primary contractor who is responsible for the worksite typically establishes the minimum requirements, controls access to the worksite, and verifies that subcontractors fulfill their health and safety duties and responsibilities. When these specifications are defined, all contractors and subcontractors should meet or exceed these requirements, as appropriate. This could be based on the nature of the assigned tasks and associated hazards [2].

In many cases, there may be several prime contractors who have responsibility for various site activities and worksite control. For example, prime contractors include the management and oversight (M&O) contractor, the construction contractor, the environmental remediation management contractor (ERMC), and site characterization and remedial design contractors. In some cases, the facility may have oversight control for all prime contractors. For example, at DOE sites, DOE has oversight responsibility for all prime contractors. In some cases, the M&O contractor also has oversight responsibility. In other cases, the M&O contractor is contractually excluded from an oversight role [3].

The responsibilities of contractors and subcontractors have been the subject of much debate. Therefore, it has become more commonplace for clients, who may have in-house personnel and resources adequate to perform cleanups, to hire subcontractors. These subcontractors could include all of those mentioned in the previous example, along with an oversight contractor. All of these subcontractors could bill the client directly for services or could bill the oversight contractor. Typically, billing the client directly can save money regarding insurance premiums or carrying charges and forces the host client to become more involved with the responsibility of running the project. This can sometimes be called a *wrap-around* in which all personnel on site are working under the same insurance umbrella.

Let's discuss an example: For time and material, or not to exceed jobs, some clients prefer to hire a general or main contractor for environmental remediation projects, and allow the contractors freedom in the performance of the work. In this case, the main environmental remediation contractor might hire subcontractors such as earth movers, haulers, reclaimers, drillers, or construction companies to perform different phases of a job. The main contractor organizes the work. Most of the work is performed by subcontractors. Once the job is completed, the client will get one invoice.

The invoice would include client services that typically include all subcontractor charges. The subcontractor charges would typically include a carrying charge, or premium, that can have a wide range. This range can start at possibly less than 10 percent and go as high as 50 percent or higher. This arrangement has advantages in that the client has little involvement with the work. If difficulties arise, they are usually resolved by the main contractor. This arrangement occurs when the client (let's say a widget manufacturer) has little or no expertise in a field (such as environmental remediation), but needs to get certain work done. The client hires the main contractor to be the "expert" for the project.

There are also disadvantages to this type of relationship. It can be costly to the client as mentioned above. An unscrupulous or irresponsible contractor may try to take advantage of the unsophisticated client. The unsophisticated client might accept responsibility that a sophisticated or experienced client might not accept. Legal action may take place.

One way to minimize these types of difficulties would be to discuss and document responsibilities before awarding a contract. Also, have a knowledgeable person or expert write the contracts. It is prudent to state relationships in contractual agreements and communicate them to all affected parties.

No matter if subcontractors are working for a general or main contractor or working directly for the client, when two or more prime contractors conduct activities at the same worksite, it is prudent that a common basis for health and safety rules and controls be established. When one contractor performs an intrusive activity that increases the hazard level for all workers at a worksite, that information should be communicated to other contractors to permit them to plan and control their activities accordingly.

Let's take another example. Under DOE, when the M&O contractor has oversight responsibility for other prime contractors, the M&O

contractor is to make certain that other contractors observe the performance standard established for the worksite and that activities are appropriately coordinated among various contractors and subcontractors. If the M&O contractor does not have oversight authority, the DOE field office assumes that function. Similar situations can often exist when working on sites where the Army Corps of Engineers has responsibility for oversight.

As outlined in the DOE requirements, the following encourage coordination and consistency among contractors:

- All contractors should interface with each other to encourage mutual understanding and coordinating their respective activities, as well as for reviewing and commenting on documents such as work plans or the safety plan.
- To make sure that all contractors and subcontractors maintain a minimum level of safety performance, the client, or general or main contractor, should establish standards for compliance. During the project planning stage, affected prime contractors should have an opportunity to provide input and resolve differences. "Cross-cut" committees are encouraged to allow prime contractors to standardize or normalize such essential elements as procedures, permit systems, and training.
- Program management and oversight contractors should establish a structure to coordinate and integrate work activities. Establishing a committee at the field level to participate in planning and overseeing can be advantageous [3].

3.4 PROJECT TEAM ORGANIZATION

The size of the project team depends on the particular tasks that are to be performed and the hazards that may be encountered. Keep in mind that a wide variety of disciplines may not be required for every project. During the early stages of planning, an organizational chart should be developed. This chart can serve to visually depict the following:

- The project team organization. This will help to identify key individuals and alternates, roles and responsibilities, and other on-site and off-site resources.
- The lines of authority, responsibility, and communication.

The organizational chart further identifies key positions in the project team, including the project director, project manager, site safety and health officer (SSHO), site supervisor, emergency response coordinator, site security, and other specialized positions.

HAZWOPER specifically requires that project personnel and responsibilities be defined [4]. Although every worker on site is expected to have responsibility regarding his or her own safety, the site hierarchy for safety-related issues should be spelled out—in other words, what procedures should be followed when a worker recognizes a safety-related situation that they cannot "fix" themselves. The plan should stress good safety principles, best management practices (BMP), and safe work behavior. Workers should never attempt to perform work for which they are not qualified. This point cannot be stressed enough. Too many times incident investigation reports will determine that the lack of trained and qualified workers was a root cause or underlying factor in a serious incident.

It is common practice for the same person to wear many hats for smaller, less complicated sites. One person could conceivably have the responsibility for the following jobs:

- Project manager
- Site supervisor
- SSHO
- QC person
- Sampling technician

On the other hand, on larger, more complicated sites, one person might have only one job, or one piece of a job. As described earlier, the SSHO may have many levels of competence. We previously mentioned three levels commonly accepted at government sites. There may be three levels or more of SSHO. Each site may be different, yet the principles are the same. The HASP should show how safety issues are addressed. Typically, a flow chart can be used to clearly depict levels of responsibility; along with straight-line versus dotted-line levels of reporting.

The following sections describe roles and responsibilities that may be included in a project team.

3.4.1 Project Manager

The project manager (PM) is typically responsible for making sure that the necessary personnel are available for the project and that the reporting, scheduling, and budgetary obligations are met.

The PM is probably ultimately responsible to make sure that all project activities are completed in accordance with requirements as outlined in the HASP. In some cases the PM may be required to perform at least one on-site safety review during the project. The PM is also responsible for making sure that all incidents are reported promptly and thoroughly investigated. The PM should approve any addenda or modifications of the HASP.

3.4.2 Site Manager

The site manager (SM) is typically the on-site representative and is responsible for maintaining contact with the host (client, customer, etc.), the PM, and the health and safety manager (HSM). The SM is also responsible for implementation of the HASP. The SM reports to the PM and works directly with the client in most cases.

The SM position will usually have some minimum qualifications. The SM should be competent, experienced, and knowledgeable in the field of specific activities anticipated during the project. If the site is a HAZWOPER site, the SM should have completed an 8-hour supervisor course as required by 29 CFR 1926.65 or 1910.120 in addition to complying with other site requirements [4]. Other responsibilities may include:

- Enforcing the requirements of the HASP. This may include performing daily safety inspections of the worksite.
- Stopping work as required to ensure personal safety and protection of property, or where life- or property-threatening noncompliance with safety requirements is detected.
- Determining and posting routes to medical facilities, emergency telephone numbers, and arranging emergency transportation to medical facilities.
- Notifying local public emergency offices of the nature of the site operations, and posting of their telephone numbers in an appropriate location.
- Observing on-site project personnel for signs of chemical or physical trauma.
- Making sure that all site personnel have been provided the proper medical clearance and have met appropriate training requirements with the appropriate training documentation, and monitoring all team members to make sure that they are in compliance with the site-specific HASP.

3.4.3 Site Health and Safety Officer

In some cases there no set requirements for a site safety and health officer (SSHO). Under the DOE there are usually specific requirements outlined in the scope of work. However, assigning an SSHO who does not have extensive training, experience, or "seasoning" can have a negative effect on the site's safety performance and safety culture. Even though there are no special courses or a set amount of field experience required, management should carefully consider the requirements set for the SSHO.

The SSHO will usually conduct daily inspections to determine if operations are being conducted in accordance with the HASP, other host contract requirements, and OSHA regulations. The SSHO is assigned to the PM for the duration of the project, but reports directly to the HSM with operational issues. An open dialogue is kept between the SSHO and supervisory personnel of the project to make sure that safety issues are quickly addressed and corrective action is taken.

The SSHO has the ultimate responsibility to stop any operation that threatens the health and safety of the team or surrounding community, or that could cause significant adverse impact to the environment. Other responsibilities include, but are not limited to:

- Implementing all safety procedures and operations on site
- Observing work team members for symptoms of exposure or stress
- Upgrading or downgrading, in coordination with the HSM and the PM, the levels of PPE based on site observations and monitoring results
- Informing the project HSM of significant changes in the site environment that require equipment or procedure changes
- Arranging for the availability of first aid and on-site emergency medical care, as necessary
- Determining evacuation routes, establishing and posting local emergency telephone numbers, and arranging emergency transportation
- Making sure that all site personnel and visitors have received the proper training and medical clearance prior to entering the site
- Establishing exclusion, contamination reduction, and support zones
- Conducting tailgate safety meetings and maintaining attendance logs and records
- Making sure that the respiratory protection program is implemented
- Making sure that decontamination procedures meet established criteria
- Making sure that there is a qualified first-aid person on site

As identified under the DOE, there are three levels of SSHO qualifications. These requirements are usually presented in job specifications and carefully outline background and experience levels required. Under the general industry and the Army Corps of Engineers you may find some other variations. The following outline will describe the three basic types of professionals.

Level 1

Level 1 sites include minimal hazards where Level D PPE is required. The minimum SSHO qualifications might include the following:

- High school education
- Work experience on projects of similar size or HAZWOPER SSHO training
- Ability to implement and verify that project activities comply with the HASP
- Current 40-hour, 8-hour refresher and 8-hour HAZWOPER training for supervisors

Level 2

For sites requiring the use of Level C PPE, the SSHO should have considerably more experience than an SSHO on a level D site. The minimum qualifications might include the following:

- Associate's degree or the equivalent in industrial hygiene, health physics, industrial safety, or other related field (work experience can be substituted if the amount and type correspond appropriately to project needs and are approved as appropriate)
- One year of health and safety work experience in hazardous waste activities that include HASP implementation
- Proficiency in use of monitoring instruments, as warranted
- Current 40-hour, 8-hour refresher, and 8-hour HAZWOPER training for supervisors

Level 3

An even more experienced SSHO should be on site when sites requiring the use of Level A or B PPE may be required. The minimum qualifications might include the following:

- Certification or eligibility for certification in industrial hygiene, safety, health physics, or related field (can substitute work experience if amount and type correspond appropriately to project requirements and are approved as appropriate)
- Two years of health and safety field experience, including hazardous waste operations, or equivalent, and demonstrated ability to implement a HASP
- Proficiency in use of monitoring instruments, as warranted
- Current 40-hour, 8-hour refresher, and 8-hour HAZWOPER training for supervisors

In addition, any SSHO designated to provide first aid or cardiopulmonary resuscitation (CPR) should meet the provision of 29 CFR 1910.1030, "Bloodborne Pathogens."

It is customary, but not required, for the SSHO to be a health and safety professional. Depending on the nature of the hazards and activities, the SSHO may be a safety professional, industrial hygienist, health physicist, engineer, health and safety technician, or even a worker with sufficient and appropriate experience and training to fulfill the established responsibilities of the SSHO (e.g., to recognize and control hazards) [3].

In more recent times, job specifications require that the SSHO report to a position that is removed from the management of the site, or at least to someone other than the SM or PM. Many times the SSHO will report to the HSM and the HSM might report to a high-ranking company official.

Selection of the SSHO is based on skills and experience proportionate to the hazards and difficulties of the job. Additional support staff can be matrixes to support the SSHO in the technical safety disciplines in accordance with project size and the nature of hazards encountered.

3.4.4 Health and Safety Manager

The HSM is typically responsible for the development, implementation, and oversight of the health and safety program. In many cases the HSM will have a minimum of three years of working experience in developing and implementing health and safety programs at hazardous waste sites. He or she should have knowledge of air monitoring techniques, development of PPE programs for working in potentially toxic atmospheres, and should have working knowledge of applicable federal, state, and local occupational health and safety regulations. The HSM will oversee and review the site operations and review and approve the HASP and any of its amendments. He or she will have a formal education and training in occupational health and safety or a related field and certification in safety management or industrial hygiene. The HSM will typically visit the site monthly or more or less as required to audit the effectiveness of this HASP, and whenever necessary to investigate major incidents.

3.4.5 Subcontractors, Visitors, and Other On-Site Personnel

Subcontractors are responsible for the health and safety of their employees and for complying with the requirements established in the HASP and the guidelines established in Safety Rules for Contractors. Subcontractors will report to the SM.

Specialty duties are assigned to teams formed for specific tasks or responding to unusual circumstances (e.g., waste characterization, confined-space rescue, asbestos, lead abatement, etc.). These teams are formed, as necessary, on a permanent or temporary basis. In many cases, special training, drills and exercises, and development of safe work plans are needed to prepare team members to conduct work safely and effectively.

For smaller projects, the field team leader and project manager will likely be the same person.

3.4.6 Occupational Physician

The occupational physician for a project should be identified and, for HAZWOPER jobs, is required to be board certified in occupational medicine [4]. For any job that involves exposure to hazardous substances, it is important that you locate and use an occupational physician (sometimes referred to in the field as an Oc Doc) who is knowledgeable about the hazards that your workers are exposed to. Sometimes in medicine, as in many other fields, working with a physician who specializes in the hazard that your company deals with can be a lifesaver.

3.5 COMMUNICATION

Communications and emergency assistance duties include the following elements: maintaining communication with work teams, assisting support zone activities, notifying emergency responders, and assisting with emergencies. In many cases these functions are assigned to a site supervisor, the field team leader, or other project team member with appropriate knowledge and experience [1].

3.6 SECURITY ISSUES

Security issues involving access controls are typically line management responsibilities. However, it is not unheard of to have the SSHO in charge of site access or other security-related matters. However, if the field team leader or site supervisors are in charge of access issues or other security matters, they should always strive for SSHO participation. The nature of a project may warrant assigning a member of the site security staff to the project team. Key duties of the security officer may include the following:

- Conducting routine area patrols
- Controlling facility access and egress
- Assisting with communication during an emergency
- Securing incident scenes
- Maintaining a log of access and egress to the worksite [1]

3.7 HAZARD CHARACTERIZATION AND EXPOSURE ASSESSMENT

The following are important aspects of a hazard-based health and safety planning process:

- Hazard characterization
- Exposure assessment and access
- Hazard controls

The amount and type of hazards will determine the performance standard specified in site-specific control plans. This includes the content, detail, and formality of review. The approval of the plans is based on risk and hazard potential. Using the hazard-based approach, levels of risk or methods to rank risk (degree) are standardized.

Professional judgment should be exercised when planning site activities and to document decisions. HAZWOPER is a performance-based standard that emphasizes hazard analyses at all stages. It encourages the development of programs that match the anticipated risk for each work activity. For example, professional judgment is used to decide if a comprehensive HASP or a scaled-down version is required for activities with little possibility to cause significant exposure.

Key documents that are developed during the planning stages can be used to focus and direct the compliance strategy, to outline the health and safety program/plan requirements, and to establish work controls. These documents are usually developed after contract award and before mobilization.

To be successful, a team selected from different groups within the project team should participate in the preparation and review of these plans. In addition, a schedule of the review and approval process for these plans needs to be established, accepted by all reviewers, and distributed before release of the first draft. Reviewers should meet an established schedule for review and submission of comments. A distinction between "review" and "approval" authority should be determined. The review process determines if all of the required elements are identified. This review can take considerable time, depending on the complexity of the project. The approval process is important when all comments are incorporated. The key responsible persons associated with the project should accept and approve the changes to the document.

Once adopted, plans should be periodically reviewed (depending on the project) and evaluated for effectiveness and cost/benefit. If the scope of work or any worksite hazards change significantly or if lessons learned indicate a review, the plans should be revised promptly.

3.8 WORK PLAN

An in-depth and detailed work plan required by 29 CFR 1910.120 (b)(3) is based on information gathered during the design phase of a project. Key planning documents are considered prestart submittals and include the comprehensive work plan, decommissioning plan, health and safety program and/or safety plan, emergency plan, and work control system (including the access and hazard controls.) It provides details on the scope of work and associated tasks, the resources required to complete the project, and the schedule. The work plan should contain the following key elements:

- Personnel requirements for implementing the work plan
- Training requirements and implementation of required informational programs per CFR 1910.120 (i)
- Identification of anticipated cleanup activities and standard operating procedures; if standard operating procedures are provided elsewhere, they are referenced and not repeated
- Defined work tasks and objectives and identification of methods for accomplishing tasks and objectives
- Provisions for implementation of the medical surveillance program
- Specialized equipment or services (for example, drilling equipment, heavy equipment operations) [3]

3.9 USING LESSONS LEARNED

Lessons learned provide valuable information for managing health and safety programs. This information addresses conditions to be avoided or recommended practices. Lessons learned typically have the potential for wide-ranging application. Effective identification of lessons learned requires an awareness of emerging practices, programs, and technologies related to hazardous waste activities [3].

The "safety alert" concept is another tool that can be used by large or small businesses to communicate past best practices and indicate a path forward. On a daily basis, lessons learned should be communicated in a site safety meeting. A worker or supervisor may have discovered that a current practice could cause a potentially dangerous situation. Many times the correction or long-term fix for a hazard involves engineering control. Therefore, if there is no imminent danger, in the short term, as work continues, communication of the hazard should take place.

The safety alert concept may also have a downside. Let's say that after an incident, the company management agrees that parts of the investigation and incident circumstances should be communicated to all workers in the company. After all, if we communicate the existence of a potential danger, we should be able to eliminate the injury in the future. To some, this is just common sense and can be considered a "no brainer." However, at a later date, should another employee of that same company suffer a similar injury, what do you think the outcome might be? The injured worker (or his lawyer) might be able to prove that the company was negligent because it knew of the problem (as shown in the safety alert) but failed to adequately address it.

There are at least a couple of conclusions we should draw from this example. If you are going to use safety alerts, even though your intentions are honorable, you may offend the injured party, the folks who took part in the investigation, and others. Be prepared for the fall-out. Also, keep in mind that you need to "talk the talk" and "walk the walk." If an incident occurs and a corrective action is indicated, it behooves you to implement some level of corrective action. If you do not implement a sound corrective action, your company likely has an ineffective safety program to go along with a variety of outstanding lawsuits.

In the DOE environment, the term "lesson learned" is defined as a "good work practice or innovative approach that is captured and shared to promote application. It may also be an adverse work practice or experience that is captured and shared to avoid recurrence." This term is used by DOE and other federal and private-sector institutions, to describe the following:

- Work processes or health and safety issues that have arisen from work at a particular site that could affect other sites or projects
- Significant experiences (both positive and negative) documented or communicated so that potentially affected operations could make changes to management practices or the conduct of operations, eliminating the hazard or helping with control of the hazard
- Lessons, problems, discussions, or potential solutions that appear in searchable databases

Exactly what type of lessons are learned cannot be foreseen. The size and diversity of site activities give rise to a wide variety of health and safety hazards. Individual sites need to document and disseminate information that could enhance their hazard recognition and mitigation. Effective documentation is an important concept that everyone needs to buy in to if the safety program is going to be effective. Why workers fail to document potential problems has been the subject of much debate.

We will not comment on why workers fail to report, but continue to believe that documentation of potential problems, unsafe conditions, and especially near misses (or more accurately referred to as "near hits") are important in the prevention of incidents at all types of sites.

3.10 CLIENT REVIEW

We must not forget the client. The client review is an excellent tool that can be used to get the client's first impression of safety performance. This has been shown to be especially effective when conducted on a formal basis after a phase of the project, or the entire project, has been completed. The PM and the SM should arrange to meet the client representative(s) in a face-to-face meeting to discuss safety performance and possibly other parameters of the recently completed job, or phase of job. The information obtained from the review is immediately analyzed. Once analyzed, it can be used to prevent recurrences of identified problems, to publicize good practices and innovative approaches to problem solving, and to perform work more safely and efficiently.

REFERENCES

1. *Occupational Safety and Health Guidance Manual for Hazardous Waste Site Activities.* Prepared by National Institute for Occupational Safety and Health (NIOSH), Occupational Safety and Health Administration (OSHA), U.S. Coast Guard (USCG), U.S. Environmental Protection Agency (EPA), October 1985, pp. 3-1, 3-4.
2. *Hazards Ahead: Managing Cleanup Worker Health and Safety at the Nuclear Weapons Complex.* U.S. Congress Office of Technology Assessment. Washington, DC: U.S. Government Printing Office, 1993, p. 7.
3. *Handbook for Occupational Health and Safety During Hazardous Waste Activities.* Office of Environmental, Safety and Health Office of Environmental Management, 1996, pp. 3-1, 3-3-3-10, 3-13.
4. 29 CFR 1910.120.

Chapter 4

Conducting a Job Hazard Analysis

You may have heard other terms such as job safety analysis (JSA), activity hazard analysis (AHA), or task-specific hazards analysis (THA). No matter what you call the term, a job hazard analysis (JHA) is a process that can used to help develop safe work practices or procedures.

A JHA is a written procedure that you can use to review job methods and uncover hazards that may have been overlooked during initial task design, process changes, and the like. A JHA is a systematic method of identifying jobs and tasks, a way of pinpointing their associated hazards, and developing procedures that will help reduce or eliminate identified risks. You can also use JHAs to document changes in a workplace and provide consistent training.

Some hazards are obvious and you can uncover them during safety reviews. Other hazards are less obvious and you can only uncover them by conducting a systematic analysis of each job to identify potential hazards.

4.1 WHY DOES A JOB HAZARD ANALYSIS WORK?

JHAs allow managers and employees to identify risks together. The manager works with the employee to record each step of the job as it is performed, consulting with the employee to identify any hazards involved in each step, and enlisting the employee's help in eliminating any hazards noted. When you develop a JHA collectively, you create a sense of ownership, thereby encouraging teamwork between the manager and the employee. This systematic gathering of information and teamwork is essential to avoid snap judgments.

Benefits of a JHA go beyond safety. As was noted in the OSHA model, the results from your JHA can and often do lead to areas such as training. Don't be surprised when your results yield an adjustment in your training program or training course content. The

JHA provides actual step-by-step safety procedures for performing each task.

Another underlying benefit for developing a JHA is providing a consistent message for new employees on a specific task or for seasoned employees who need safety awareness training or review of their specific task.

In addition, a properly designed JHA is a good learning tool that you can use to evaluate incidents. Job-related incidents occur every day in the workplace. These incidents, which include injuries and fatalities, often occur because employees are not trained in the proper job procedures. One way to reduce these workplace incidents is to develop proper job procedures and train all employees in the safer and more efficient work methods.

The JHA allows you to identify weak links in the system. Once you discover the weak links, you can update the JHA to reflect the needed changes.

Let's consider some of the important cost factors of a JHA. These methods can help to improve job procedures and can help to reduce costs that result from absenteeism and workers' compensation claims, as well as hidden costs that are usually overlooked. These hidden costs include management time for investigation; lost time for other workers who experience some level of trauma; hiring and training temporary workers; bad publicity, poor product quality, employee morale; OSHA citation/fines, court costs, and so on. Reduction of these costs can lead to increased productivity and improved cost to the bottom line.

Establishing clear job procedures is one of the benefits of conducting a JHA, carefully reviewing and recording each step of a job or related task that make up the job, identifying existing or potential job safety and health hazards, and determining the best method to perform the job or to minimize or eliminate the associated hazards.

There is one major drawback. A JHA program takes time, both to document and to implement effectively, and is a continuous improvement tool that is forever changing. However, as you will see, the positive benefits outweigh the time required.

The JHA is not a mandatory requirement or a standard, and you are not required to use the recommended methods. It is considered a management tool and a BMP, going beyond the OSHA standard.

4.2 SELECTING THE JOBS FOR ANALYSIS

You should conduct a JHA for all jobs or tasks in a workplace, no matter if the job or task is existing, new, routine, nonroutine, or needs special consideration. You should consider even one-step jobs, such as those in which only a button is pressed. They should be analyzed by evaluating surrounding work conditions.

To evaluate a job effectively, you should have some experience and be trained in the intended purpose of the JHA, have an open mind, and have examples of correct methods. Focusing on safety is essential to the job being evaluated.

To determine which jobs you should analyze first, review your injury and illness reports such as the OSHA 200 log, your medical case histories, your first-aid cases, and workers' compensation claims. First, you should conduct a JHA for jobs with the highest rates of disabling injuries and illnesses. Do not forget jobs in which you have had "close calls" or "near hits." You should give these incidents a high priority. Analyses of new jobs and jobs in which changes have been made in processes and procedures should be the next priority.

In addition, when selecting the job for analysis the following points can be useful in setting priorities:

- Injury and occupational illness severity. Those jobs that have involved serious incidents. There may be a basic problem in the work environment or in the job performance itself.
- Accident frequency. The higher the frequency rate of incidents, the greater the reason for implementing a JHA.
- Potential for illness or injury, even if no such incident has occurred.
- A new job or task with no accident history or information about its potential for incidents. Many incidents occur in a job or task where the employee is not accustomed to the job.

To be effective, the creation of a task or modification of a task through the introduction of new processes or equipment should automatically require you to develop a new or revised JHA. Jobs with many steps are usually good candidates. As stated before, you should assign each job selected a priority based on the accident potential and the severity of associated potential injuries.

4.3 EMPLOYEE PARTICIPATION

Once you have selected the job for analysis, discuss the procedure with the employee who performs the job and explain the intended purpose. Point out that you are studying the job itself and not checking on the employee's job performance. Involve the employee in all phases of the analysis, from reviewing the job steps and procedures to discussing potential hazards and recommended solutions. You should also talk to other employees who have performed the same job in the past.

Employees are the best source for identifying job hazards, and they appreciate you consulting with them on matters that affect them. Employees become more receptive to changes in their job procedures when you give them an opportunity to help develop the change.

4.4 CONDUCTING A JHA

It is important to remember that direct management support should be available. Without this support a good analysis will likely not be conducted.

Once the hazards have been identified, the correct solutions can be developed to protect the employee from physical harm. Once the jobs have been selected, determine how the JHA will be conducted. Two methods can be used to begin the analysis: the discussion method and the observation method.

Discussion Method

This is the simplest and least expensive method. The manager will sit down with the employee and discuss the JHA. Only obvious hazards are identified in this initial session. These observations are based on the recollections and observations of employees who have performed the job. This information is valuable because it relies on the experience of employees closely linked to the job.

Observation Method

This method involves going to the job location and observing the tasks as they are completed. The employee is interviewed about the hazards inherent in each task. This method is better than the discussion method but has some drawbacks. You are limited by your powers of observation.

Before beginning the JHA, observe the general work area. Since each job involves a different sequence of activity, you should observe how the job is performed. Then you should develop a checklist. The following list shows some sample questions you might ask.

- Are there materials on the floor that could cause a tripping hazard?
- Is there adequate lighting?
- Are there any live electrical hazards?
- Are there any chemical, physical, biological, or radiation hazards associated with the job? Are any of these hazards likely to develop?
- Are tools, including hand tools, machines, and equipment, in need of repair?

- Is there excessive noise that may hinder communication or is likely to cause hearing loss?
- Are job procedures understood and followed and modified as applicable?
- Are emergency exits clearly marked?
- Are industrial trucks or motorized vehicles properly equipped with brakes, overhead guards, backup signals, horns, steering gear, seat belts, etc.? Are they properly maintained?
- Are all employees who operate vehicles and equipment authorized and properly trained?
- Are employees wearing proper personal protective equipment (PPE)?
- Have any employees complained of headaches, breathing problems, dizziness, or strong odors?
- Have tests been made for oxygen deficiency, toxic vapors, or flammable materials in confined spaces before entry? Is ventilation adequate, especially in confined or enclosed spaces?
- Are workstations and tools designed to prevent twisting motions?
- Are employees trained in what to do in the event of a fire, explosion, or toxic gas release?

This list is only a sample of some of the hazards that you may encounter when conducting a JHA. The list is by no means complete. Each worksite may have its own unique requirements and environmental conditions. You should add your own questions to the list.

4.5 BREAKING DOWN THE JOB

There are fundamental issues that should be considered when developing JHAs:

- Select a capable person to review jobs
- Train the person in the proper techniques of conducting a JHA
- Observe the employee doing the job and ask for the employee's input
- Record each step of the job or task
- Verify to make sure that all job steps have been identified
- Review the steps in which hazards exist.

You can break down every job into basic tasks or steps. To begin, list each step in the order of occurrence as you watch the employee perform the job. No basic step should be omitted. Make sure you record enough information to describe each action. When this is completed, review the steps with the employee.

You should be careful not to omit any steps. Care should be taken not to make the job hazards too detailed. Too much detail will make a JHA ineffective. Make sure that only "safety steps" are recorded. One of the common mistakes is to mix work elements with job hazards. A JHA is not intended to document work process instructions, although some people believe that they should be included.

Talk to as many people as possible: new, experienced, transferred, and temporary employees, managers, maintenance personnel, safety professionals, and so on. Common problems will soon become apparent. Not only will you base your decision on better information, but also people will react favorably at having been consulted. Discuss potential solutions with technical specialists and with employees.

4.6 IDENTIFYING JOB HAZARDS

After you record the steps of the job, review each step to determine the hazards that exist or that might occur. There are several ways to identify job hazards: evaluate the ways human error might contribute to a hazard, record the types of potential incidents and the physical agents involved, and make sure that procedures are clearly written.

Once the jobs have been identified and the basic steps outlined, the hazards can be identified. Evaluate each step as often as possible to identify all real hazards. Both physical and mechanical hazards should be considered. Review the actions and positions of the employees. Ask yourself these kinds of questions:

- Is the employee wearing PPE?
- Are work positions, machinery, pits or holes, and/or hazardous operations adequately guarded?
- Are lockout procedures used for machinery deactivation during maintenance?
- Are there fixed objects that may cause injury, such as sharp edges on equipment?
- Is the flow of work properly organized (i.e., is the employee required to make movements that are rapid)?
- Can reaching over moving machinery parts or materials injure the employee?
- Is the employee at any time in an off-balance position?
- Is the employee positioned at a machine in a way that is potentially dangerous?

- Is the employee required to make movements that could lead to or cause hand or foot injuries, strain from lifting, or repetitive motion injuries?
- Do environmental hazards such as dust, chemicals, radiation, welding rays, heat, or excessive noise result from the performance of the job?
- Is there danger of striking against, being struck by, or contacting a harmful object? Employees can be injured if they are forcefully struck by an object or contact a harmful material.
- Can employees be caught in, on, by, or between objects? Employees can be injured if their bodies or part of their clothing or equipment is caught on an object that is either stationary or moving. They can be pinched, crushed, or caught between either a moving object and a stationary object, or two moving objects.
- Is there a potential for a slip, trip, or fall? Can employees fall from the same level or a different level?
- Can employees strain themselves by pushing, pulling, lifting, bending, or twisting? Employees can also overextend or strain themselves while doing a task and strain their backs by twisting and bending.

Note equipment that is difficult to operate and could be used incorrectly. Make sure that all equipment is in proper working condition. Determine what stress level the employee is experiencing.

What other hazards not discussed have the potential to cause an incident? Repeat the job observations as often as necessary until all hazards have been identified.

4.7 RECOMMENDING SAFE PROCEDURES AND PROTECTION

After you have generated a list of hazards or potential hazards and have reviewed them with the employee, determine if the employee can perform the job another way to eliminate the hazards, such as combining steps or changing the sequence. You should be aware if safety equipment and precautions are needed to control the hazards.

If safer and better job methods can be used, list each new step, such as describing a new method for disposing of material. List exactly, as you would in a training objective, what the employee needs to know to perform the job using a new method. Do not make general statements about the procedure, such as "be careful." Be as specific as you can in your recommendations. You may wish to set up a training program using the JHA to retrain your employees in the new procedures, especially if they are working with highly toxic substances or in hazardous situations. (Some OSHA standards require that a formal training program should be established for employees.)

If you cannot develop a new procedure, try to determine if any physical changes could help to eliminate or reduce the danger. These changes may include redesigning equipment, changing tools, or adding machine guards, PPE, or ventilation.

If hazards are still present, try to reduce the necessity for performing the job or the frequency of performing it. Go over the recommendations with all employees performing the job. Their ideas about the hazards and proposed recommendations are valuable. Be sure that they understand what they are required to do and the reasons for the changes in the job procedures.

4.8 REVISING THE JHA

JHAs can do much toward reducing incidents in the workplace. The JHA is only effective if you review and update it periodically. Even if there are no changes in a job, you may detect another hazard that was missed in an earlier analysis.

If an incident does occur, you should review the JHA immediately to determine if changes are needed in the job procedure. In addition, if a "close call" or "near hit" has resulted from an employee's failure to follow job procedures, you should discuss these incidents with all employees performing the job.

Any time you revise a JHA, employees affected by the change should be trained in the new job methods, procedures, or protective measures. A JHA also can be used to train new or transferred employees in the basic job steps and associated hazards.

To show how a JHA form is prepared, a sample worksheet for cleaning the inside of a chemical mix tank is provided in Appendix D. Both safety and health hazards are noted, as well as recommendations for safer methods.

4.9 PROCESS HAZARD ANALYSIS

Included in OSHA's JHA Booklet, 3071, is a good description of a process hazard analysis (PHA) [1]. This is being used in the Process Safety Management (PSM) program (29 CFR 1910.119) to understand how hazards exist. There are some good methods listed in the manual that can be used to conduct a JHA. As you review each method you can determine which one may be useful for your operation. The typical method chosen is the checklist.

PSM was created to help the management of hazards associated with processes using highly hazardous chemicals. In an appendix to the rule, OSHA discussed several methods of process hazard analysis. That discussion, which may be helpful for those doing job hazard analyses, follows.

What If?

For a relatively uncomplicated process, review the process from raw materials to finished product. At each handling or processing step, you formulate and answer "what-if" questions to evaluate the effects of component failures or procedural errors on the process.

Checklist

For more complex processes, you best organize the "what if" study through the use of a "checklist," and the assignment of certain aspects of the process to committee members having the greatest experience or skill in evaluating those aspects. Operator practices and job knowledge are audited in the field, the suitability of equipment and materials of construction is studied, the chemistry of the process and the control systems are reviewed, and the operating and maintenance records are audited. Generally, a checklist evaluation of a process precedes use of the more sophisticated methods described below, unless you have operated the process safely for many years and the process has been subjected to periodic and thorough safety inspections and audits.

What If/Checklist

The what if/checklist is a broadly based hazard assessment technique that combines the creative thinking of a selected team of specialists with the methodical focus of a prepared checklist. The result is a comprehensive hazard analysis that is useful in training operating personnel on the hazards of the particular operation.

The review team is selected to represent a wide range of disciplines, such as production, mechanical, technical, and safety. Each person is given a basic information package regarding the operation to be studied. This package typically includes information on hazards of materials, process technology, procedures, equipment design, instrumentation control, incident experience, and previous hazard reviews. A field tour of the operation is also conducted. The review team methodically examines the operation from receipt of raw materials to delivery of the finished product to the customer's site. At each step, the group collectively generates a listing of "what-if" questions regarding the hazards and safety of the operation. When the review team has completed listing its spontaneously generated questions, it systematically goes through a prepared checklist to stimulate additional questions.

Subsequently, the review team develops answers for each question. They then work to achieve a consensus on each question and answer. From these answers, a listing of recommendations is developed specifying the need for additional action or study. The recommendations, along with the list of questions and answers, become the key elements of the hazard assessment report.

Hazard and Operability Study (HAZOP)

HAZOP is a formally structured method of systematically investigating each element of a system for all ways where important parameters can deviate from the intended design conditions to create hazards and operability problems. The HAZOP problems are typically determined by a study of the piping and instrument diagrams (or plant model) by a team of personnel who critically analyze effects of potential problems arising in each pipeline and each vessel of the operation.

Pertinent parameters are selected, for example, flow, temperature, pressure, and time. Then the effect of deviations from the design conditions of each parameter is examined. A list of keywords, such as "more of," "less of," "part of," is selected for use in describing each potential deviation.

The system is evaluated as designed and with deviations noted. All causes of failure are identified. Existing safeguards and protection are identified. An assessment is made weighing the consequences, causes, and protection requirements involved.

Failure Mode and Effect Analysis (FMEA)

The FMEA is a methodical study of component failures. This review starts with a diagram of the operations, and includes all components that could fail and conceivably affect the safety of the operation. Typical examples of components that fail are instrument transmitters, controllers, valves, pumps, and rotometers. These components are listed on a data tabulation sheet and individually analyzed for the following:

- Potential mode of failure (i.e., open, closed, on, off, leaks)
- Consequence of the failure; effect on other components and effects on whole system. Hazard class (i.e., high, moderate, low)
- Probability of failure
- Detection methods
- Compensating provision/remarks

Multiple concurrent failures are also included in the analysis. The last step in the analysis is to analyze the data for each component or multiple component failure and develop a series of recommendations appropriate to risk management.

Fault Tree Analysis (FTA)

An FTA can be either a qualitative or a quantitative model of all the undesirable outcomes, such as a toxic gas release or explosion, which could result from a specific initiating event. It begins with a graphic representation (using logic symbols) of all possible sequences of events that could result in an incident. The resulting diagram looks like a tree with many branches. The diagram lists the sequential events (failures) for different independent paths to the top or undesired event. Probabilities (using failure rate data) are Process Hazard Analysis assigned to each event and then used to calculate the probability of occurrence of the undesired event.

The technique is particularly useful in evaluating the effect of alternative actions on reducing the probability of occurrence of the undesired event.

4.10 SUMMARY

A JHA documents procedures that can be used to review job methods and uncover hazards that may exist in the workplace. JHAs can also be used to document changes in work tasks. Some solutions to potential hazards may be physical changes that eliminate or control the hazard or a modified job procedure that will help eliminate or minimize the hazard.

All employees should be trained in how to use the JHA. Managers are in the best position to do the training by observing the job as it is being performed to determine whether or not the employee is doing the job in accordance with the job procedures.

A JHA should be monitored to determine its effectiveness in reducing or eliminating hazards. You should also find out whether the employee is following the analysis when performing the job. If so, evaluate the effectiveness. If not, try to find out the reason.

It is important to assign both authority and specific responsibility to implement each protective measure. A safety engineer may need to provide the training; the manager should provide safe tools and equipment; and the employees should inspect their tools to ensure that they are in safe condition.

Everyone has seen the demonstration in which you start off by telling a story to the first person in a group. The story is then passed on to the next person, and so on down the line. By the time the story gets back to the original storyteller, the message has changed. In this case, if they had a written script similar to a JHA, then the story would have been the same message around the room.

We need to remember that JHAs should be easily readable and that the hazards need to be easily understood. For readability, JHAs need to be typed. They should be placed at every workstation. It is important to highlight the most critical hazards for special attention. The objective is to make a JHA a user-friendly document that everyone can read to understand the hazards of the identified task [1].

REFERENCE

1. *Job Hazard Analysis*. U.S. Department of Labor, OSHA 3071, 1998 Revised, pp. 3–16.

Chapter 5

Developing a Site-Specific Health and Safety Plan

A properly designed and implemented site-specific HASP provides the basis for protection of workers, visitors, and the public. The HASP is a requirement at all HAZWOPER sites [1] and will likely soon become a requirement at all sites. However, before we begin development of the HASP there is a lot of work to do. The following discussion will outline the differences between a health and safety program and a HASP document.

The health and safety program is usually defined by a broad-based document that is often referred to as a policy and procedure (P&P) manual or accident prevention standards. These documents are general in nature and provide general guidance on how the company handles safety-related issues. Earlier we discussed how certain companies use regulatory compliance as a measuring stick, but have goals that go well beyond compliance. The safety program is the vehicle that is used to communicate the company philosophy.

The HASP, on the other hand, focuses on the site-specific activities and outlines the appropriate elements of the site's existing health and safety program to the related task. The existing programs are reviewed to identify those elements meeting the needs of the planned site activity. Program elements and procedures are supplemented with worksite-specific detail and tailored to meet special or unique aspects of the hazardous waste activity on an as-needed basis [1].

5.1 IDENTIFYING RESOURCES

The planning process also includes careful analysis of the need for and timing of resources to conduct hazardous waste activities. Resources that my be used include the following:

- Qualified personnel to fill staffing assignments. This might involve an employment agency specializing in supplying HAZWOPER qualified persons on a temporary basis.

- Equipment, facilities, supplies, tools, and utility services (e.g., PPE, sampling equipment, instrumentation, hot/cold water, electricity, sewage treatment).
- Outside support services (for example, medical surveillance; laboratory analyses; training consultants; emergency response to incidents, injuries, fires, and hazardous materials incident responders or experts as required by site activities).

Adequate resources are fundamental to any project, but even more so to a well-functioning health and safety program. Many mishaps have been traced to improperly trained workers, lack of adequate tools and equipment, or requirements for personnel to work excessive hours or at unfamiliar jobs because of inadequate staffing or lack of adequate resources. The multidisciplinary team approach can help to identify required resources and can help balance, identify, and coordinate necessary assets.

5.2 UNDERSTANDING THE SCOPE OF WORK

The author(s) of the HASP should have an in-depth understanding of the expected scope of work (SOW). One can obtain this in-depth understanding in a variety of ways, the most obvious of which is to discuss the SOW with the project manager. The PM may be able to give the author insight as to site activities or history. In addition, establishing a relationship with the PM from the beginning will make a smoother transition and better communication down the road. Keep in mind that discussing the SOW with the project manager sounds like an easy task, but is often not so easy. PMs, like most other busy people, have a tendency to be working on a variety of projects at one time. Getting a face-to-face audience with them can be difficult. In addition, PMs have been known to be "bounced around" between projects, and for that matter between employers. Just because you may have gotten a good idea about the project prior to writing a HASP, that does not mean that the PM or SOW has not changed many times.

The next step is to talk to individual task managers or others who can expect to perform the work. This may include subcontractors or other personnel who will perform work on the site. If contractors are involved, a decision should be made as to the relationship and responsibility for overall safety responsibility. If the HASP is likely to be adopted by others, this should be kept in mind during development.

Besides incentives, contracts are important because they define what is supposed to happen and who will be responsible for what. The HASP document should reflect site activities. Having a good understanding of contractual terms and including pertinent requirements in the HASP can serve to reinforce the contract.

Contract documents should be reviewed. The HASP should reflect and possibly reference contractual agreements. Contract documents can contain much information pertinent to site safety. For example, many contracts contain monetary incentives for completion of site work accident free. If management wishes to share some of this monetary incentive with site workers, the HASP is an excellent vehicle for communicating safety incentive programs.

Job specifications should be reviewed. For larger sites job specifications may be many inches thick, and for small sites job specifications may not exist. For those sites where job specifications do exist, they should be studied in detail. The specifications will typically supply the author(s) of the HASP with pertinent information about the customer requirements.

5.3 HASP PREPARATION

The HASP is the model for performing work safely and, if properly designed, will help to integrate all site- and task-related hazards and control measures. When working with the DOE or the Corps of Engineers there may be additional documents that should be considered when developing a plan. Make sure that you understand all site-specific needs.

When a worksite includes both radiological and nonradiological hazards, the HASP should address both hazards. A site-specific HASP can supplement health and safety programs by providing site-specific and pertinent information, requirements, and strategies for each task.

A properly written HASP should contain worker health and safety program information, guidance, and alternatives. The HASP should quickly answer the following questions: "What hazards are present?" and "How can we make sure that the tasks will be performed safely?" The following general guidelines will help to answer these questions:

- Each HASP should address only one worksite. Copies should be maintained at the worksite, readily accessible and placed in an appropriate location. A HASP should be developed prior to any preliminary worksite assessment. Information from this assessment can be used to modify the HASP to reflect subsequent worksite activities. In general, the HASP is based on hazard analyses and should be updated periodically to reflect the ever-changing worksite conditions and activities as the project progresses.
- The document should be designed to be concise, user friendly, and usable as a reference for both supervisors and workers. It should help identify hazards and implement hazard control requirements for the

site. Workers should be able to read the HASP and learn what hazards will be encountered and what controls are in place to mitigate them.
- If your company is working under a HASP developed by someone outside your company, there are other considerations that should be examined. If you review the HASP and you do not feel that a certain section applies to your operation, you should make sure that you document this exception in writing. Keep the documentation on file at the site and keep a backup at another location. It is important to keep this documentation on file and to communicate the changes so that all site personnel understand their safety obligations.

Before undertaking development of any HASP, some of the following also should be considered.

Does each worksite require a separate HASP, or can one plan cover multiple worksites? In general, each HASP should address only one worksite. However, this is not a requirement. A situation could arise in which it is decided to use one HASP for multiple worksites. We believe that the approach used should depend on conditions at the worksite. If worksites are similar, in near proximity to each other, and activities are phased together, one HASP may be preferred. If worksites have enough differences that need to be addressed in the HASP and these differences could cause confusion in the field, then more than one HASP should be developed.

Having a standard format for HASP development is recommended. Those people who will be the primary users will be more comfortable and willing to use the HASP in a format that they are familiar with. This ensures both comfort with using the HASP and cost-effectiveness. Constructing a single HASP template for various types of activities is generally more cost-effective than developing each HASP from scratch. The template can be tailored to site-specific conditions and activities. It is also possible to construct an "umbrella" HASP with basic requirements and guidance applicable to several different worksites, thereby streamlining the preparation process by drawing on common conditions. This approach might be appropriate for a tank farm operation comprising individual farms or tanks with distinct hazards and similar operations, for a grouping of similar facilities undergoing deactivation, or for well sampling or installation activities [1].

Two questions commonly asked by the unfamiliar manager are: "Why isn't the existing health and safety program enough? Why is a HASP needed?" The sitewide health and safety program typically includes many procedures (e.g., lockout/tagout, hearing conservation) that are referenced in the HASP and applied to the hazardous waste worksite. The safety program is general in nature and is meant to be general. In addition, although the safety program contains valuable safety information, referring to the program is not sufficient.

The HASP focuses on the specific tasks down to the worksite level and identifies job- and task-based hazards, exposure-monitoring requirements, hazard controls and approaches, requirements necessary to protect workers, and, sometimes, the name of the person responsible for a certain activity.

For example, on a site where scaffolds are being used there would be a general scaffold procedure. This procedure should be part of your organization's safety program. This general program should be part of the HASP or included as a reference. In addition to the general program part of scaffolds, you also need to talk about site-specific scaffold safety information.

The general procedure, which is part of the "program," never changes. The site-specific HASP, on the other hand, changes with each site because the site-specific information such as locations and types of scaffolds, competent persons, and training requirements will likely change. An overall health and safety program simply does not have the specificity necessary to meet all HASP requirements for a given work activity.

Not all existing procedures or program elements of the overall health and safety program need to be incorporated into the HASP. For example, if noise is a hazard, the plan does not have to cite the entire hearing conservation program. Procedures already established elsewhere may be referenced, as applicable. In another example, if a confined-space-entry procedure is required, the HASP could reference the particular procedure which is part of the overall program. The next step would be to identify confined spaces at the worksite where the procedure applies, and then provide appropriate implementation procedures (e.g., conditions to be monitored, evaluation of the space, issuance of an entry permit). If special operational procedures apply to the worksite, they can be attached to the HASP using an appendix.

Not all required tasks and hazards can be predicted during the development of a HASP. The plan describes the ongoing hazard analysis and work control process, defines the means of identifying job- or task-based requirements and controls, and discusses ways to inform workers about requirements derived from ongoing job or task hazard analyses.

Work planning and control processes include the use of job hazard analyses (JHAs), job safety analyses (JSAs), task analyses, safe work plans, safe work permits, or procedures.

Hazardous waste operations often include tasks and activities that are conducted on a periodic basis, are of very short duration, are transient in nature, or otherwise pose little hazard. Developing a brief HASP template (e.g., "fill in the blank"), a permit, or a checklist system that includes essential HAZWOPER-type information may suffice for these types of operations.

Each worksite is different even though relative hazards may be similar. Wind direction, surface contamination, or neighboring properties that may contain contributing contamination may vary. The site description/background sections should give the workers enough information to perform their job safely without overkill. The simpler, smaller, and fewer hazards there are on site, the less background information will be necessary.

5.4 HAZARD CHARACTERIZATION AND EXPOSURE

DOE sites not only are subject to radiological hazards, but also have the typical physical, chemical, and biological hazards associated with other sites. Although your site may seem unlikely to have radiological hazards, they can be found in more places than you might believe. For example, if your site manufactures chemicals or other items, or generates electricity, it likely has some form of radiological hazards.

5.4.1 Radiological Hazards

Unlike many chemical hazards, radiological hazards can be easy to detect with highly sensitive, direct reading instruments. Radiological control personnel conduct surveys and post warning signs.

The important aspect is to know how to control or limit your exposure to radiological hazards. Some of the solutions can be summarized as follows:

- **Time:** Limit the time you are exposed to a hazard.
- **Distance:** Use robotics or tools to increase your distance from the hazard.
- **Shielding:** Use dense materials as shielding or place enclosures around the source.

Anyone working with different types of radioactive material should know the conditions when various materials may be present. The following provides some additional guidance as to where radioactive materials may be present:

- Contaminated soils
- Loose, fixed, surface, subsurface, or airborne contamination
- Drums or containers of contaminated liquids and solids
- Equipment or system components generating radiation or contaminated by radioactive materials
- Activated materials

- Sealed and unsealed sources
- Areas near operating nuclear reactors [2]

5.4.2 Exposure Monitoring

Air monitoring should be developed based on an initial assessment. This should be site-specific, taking into consideration all potential site hazards. Air monitoring can vary as widely as contaminants. For site screening purposes, direct reading instruments are often used. In many cases direct reading instruments cannot be used effectively when time-weighted average (TWA) information should be obtained. There are a variety of direct reading instruments that can be used to determine the airborne concentrations of a variety of chemicals. However, when the site becomes complicated by containing a variety of hazardous materials that have varied properties such as ionization potentials, choosing direct reading instruments can become a challenge. A skilled operator is typically an asset anytime instrumentation is used. Although manufacturers make an effort to simplify instrument use, there is no substitute for time, training, and experience.

The HASP should be designed to specify action levels that would cause the worker to upgrade or downgrade PPE.

5.4.2.1 Air Monitoring

Only qualified individuals should be allowed to develop air monitoring strategies. In addition, only trained and qualified field personnel should operate screening equipment and be allowed to interpret results. For many sites, the results obtained from direct reading instruments can help determine a variety of important factors on a hazardous waste site. These factors include:

- Work zone determinations
- Respiratory protection requirements
- PPE requirements such as whole body protection
- Decontamination requirements

Direct reading instruments also can provide an indication of site emission levels.

An exposure assessment uses air-monitoring data to determine possible worker exposures. This data is used to identify controls for worker protection and provide monitoring results to physicians for proper medical assessment, treatment, and follow-up care.

Colorimetric detector tubes are commonly used when instantaneous results of individual constituents are needed immediately. However, you should be careful when using colorimetric tubes because of their limitations. Many times, interference chemicals are present that can cloud the results. Although the use of colorimetric tubes is not perfect, using them has some distinct advantages, including:

- **Simplicity:** Compared to most of the instrumentation currently in use, colorimetric tubes have no moving parts.
- **Sturdiness:** Colorimetric tubes hold up under tough conditions.
- **Reliability:** Colorimetric tubes are unaffected by power surges or outages. Tubes do not rely on electrical power. If the power goes out or surges, it does not have an effect on the operation.
- **Portability:** Tubes can easily be used in remote areas. They are light in weight and easily carried in hard-to-reach areas.
- **Wide ranges for use:** Tubes usually are not affected by high humidity or temperature.

Mobile laboratories are another alternative. If large amounts of data need to be analyzed in a short time frame, having a lab on site can be a real asset. A local lab may also provide similar service.

Field test kits have been used with success on many sites for a variety of contaminants. The types of test kits available and their continued use appear to be a wave of the future. Certainly, the skills of the sampling technician and field test kit user are two very important areas for consideration.

Air monitoring data is important because of the sensitive nature of the data collected. Data collected should be given a high priority. Air monitoring should be taken seriously. If abnormal readings (high or low) are observed, immediate action should be taken to determine accuracy. It should be decided if levels of protection need to be adjusted or if other appropriate action should be taken. All readings taken should be recorded in a logbook and become part of the site permanent record and project file. Reading results of "0" or nondetect should be recorded. After all, when it comes to screening equipment from the safety and health point of view, "0" is a very important number.

5.4.2.2 Noise Monitoring

Noise monitoring is usually located in the HASP as part of the monitoring program. Noise monitoring should be performed in accordance with acceptable practices. Typically, noise levels are monitored in the field with either a Type I or Type II sound level meter (SLM). Noise dosimeter readings can also be obtained to determine the percent (%) noise dose. Noise levels and % doses measured are then compared to limits listed in OSHA standard 29 CFR 1910.95, Hearing Conservation [3].

Noise monitoring equipment, like air monitoring equipment, should be used, stored, and maintained in accordance with manufacturer recommendations. Noise instrumentation is typically calibrated prior to use for each shift and checked at the end of the shift to determine accuracy. Noise readings also should be recorded in a log and should become part of the permanent site record.

Typically, if workers are working an eight-hour day, selection of hearing protection will match the employees' needs and the ability to attenuate noise below the required standard. If the hours an employee works are greater than eight hours, the noise attenuation levels should be adjusted accordingly. Each hearing protection device (muff or plugs) has a noise reduction rating (NRR) associated with it. There are a variety of ways to calculate the hearing protector's effectiveness. One commonly used formula is as follows: noise reading dB(A) − (NRR − 7dB) < 90 dB(A). There are a variety of other methods to calculate noise reduction. The important thing to remember is that no matter which calculation you use, the higher the number associated with the NRR, the better or more protective the equipment.

5.5 CHEMICAL HANDLING PROCEDURES

Workers should be trained in the hazards that they are potentially exposed to. The handling procedures that are adopted should ensure that whatever the hazards present, they are adequately controlled. Some typical control procedures include:

- Stand up-wind of chemical products whenever possible.
- Minimize direct contact and contact time with contaminated media.
- Avoid walking through discolored areas or puddles, leaning on drums, or contacting anything that is likely to be contaminated.
- Do not eat, drink, smoke, or apply cosmetics in the hot or warm zones.
- Wear appropriate PPE when it is necessary to come into contact with contaminated media or surfaces.

Because this list is general in nature, the user will have to modify it to be sure that site specificity and applicability are taken into account.

5.5.1 Airborne Dust

Typically, you will find that there is a reasonable concern about exposure to dusts on many sites. It you find that dusts are a potential concern, consider adding the following information to the HASP document.

- Stand up-wind whenever intrusive activities occur and generate visible signs of airborne dust.
- Monitor for airborne dust (surface or subsurface soil) with portable, aerosol dust direct reading instruments.
- Utilize wet methods (spraying ground, wet drilling, etc.) when visible signs of airborne dust are generated.

5.6 WORK ZONES

Work zones are often temporary. Many times, once the work has been conducted, the zone boundaries change and sometimes become support zones. Temporary work zones can be used to effectively manage regulatory scope. Area and personnel exposure monitoring is crucial in order to verify that zoning, containments, work practices, and procedures have been designed appropriately and maintain worker health and safety.

There are three main types of work zones at a hazardous waste site: the exclusion (or hot) zone, the contamination reduction (or decontamination) zone, and the support (or cold) zone. The following provides a discussion of each type of zone.

5.6.1 Exclusion Zone

The exclusion zone is where contamination is present and there is the highest possibility for worker exposure to hazardous materials. The HASP should specify the PPE requirements for all work conducted in this zone. Without exception, workers who enter the exclusion zone should wear specified PPE. The level of protection may vary based on activity, stage of the operation, or location. For example, most of the exclusion zone might have a relatively low exposure potential that could be controlled by Level D PPE; an area undergoing subsurface remediation in the zone might be set apart and controlled by Level B PPE; and another area might contain surface chemical contamination and require Level C PPE.

Access control points are established at the hot zone to regulate the flow of personnel and equipment into and out of the zone. Separate entrances and exits are provided for personnel and heavy equipment.

When establishing the exclusion zone, you should visually survey the worksite to review the following items:

- Determine the locations of the hazardous materials and substances: drainage, leachate, spilled material, and visible discolorations.

- Evaluate the initial direct reading instrument survey data for the presence of combustible gases, organic and inorganic gases, particulates, vapors, and ionizing radiation.
- Evaluate air, soil, and water sampling results. Consider the distances needed to prevent an explosion or fire from affecting personnel outside the exclusion zone.
- Consider the area necessary for site work to reduce the spread of contamination.
- Consider meteorological conditions and the potential for contaminants to be blown from the area.
- Secure the hot line using appropriate barriers and posting.
- Modify the hot line location, if necessary, as more information becomes available.

5.6.2 Contamination Reduction Zone/Corridor

The contamination reduction zone/corridor (CRZ/C) is where decontamination is performed and is identified as the entry and egress route between the exclusion and support zones. The CRZ/C reduces the probability that the clean area or support zone becomes contaminated and affected by site hazards by limiting the transfer of hazardous substances. The CRZ/C is positioned and maintained in a condition that requires minimal use of PPE, but decontamination workers still wear PPE appropriate to the hazard. The level of PPE required in the CRZ/C is specified in the HASP and is usually one level below the hot zone. The following outlines the CRZ/C design:

- Personnel and equipment decontamination (e.g., separate lines for workers and heavy equipment such as tractors, earth-moving equipment, trucks, and other material)
- Emergency response functions (including transport of injured personnel, first-aid equipment, and containment equipment)
- Equipment resupply
- Sample packaging and preparation for on-site or off-site laboratories
- Location of worker temporary rest areas
- Drainage of water and other liquids used in the decontamination process
- Waste minimization
- Reduction or elimination of mixed waste production

The CRZ/C's primary purpose is to keep the support zone free of contaminants and hazards. The size and location of the CRZ/C should be based on the stability of site conditions, the potential for dispersion of contaminants and for unexpected events, and the proximity of uninvolved workers and third parties. The CRZ/C boundaries are established based on hazard characterization and do not need to encircle the entire perimeter of the exclusion zone.

5.6.3 Support Zone

The support zone also is called the clean zone. This is where administrative and support functions necessary to maintain site controls are located. The support zone location should be based on the following six general criteria:

- Accessibility
- Resources
- Visibility
- Prevailing wind direction
- Distance from exclusion zone
- Type of work

Normal work clothes are appropriate for the support zone. PPE worn for the hazardous waste work should remain in the CRZ/C. At some point, this PPE will be decontaminated or packaged for transport and disposal or decontamination. Separate support zone facilities may not be needed where site facilities are readily available and near to the worksite, and if close communication is maintained. For multiple hazardous waste operations conducted in close proximity, it is possible to design one support zone to serve several operations. This will depend on the logistics of the project.

A properly equipped support zone may consist of a single trailer or may be composed of multiple facilities such as a command post, medical station, equipment and supply centers, field laboratory, and administrative areas.

The following elements should be taken into consideration when determining the location and setup of the support zone:

- Accessibility
- Proximity to highways and railroad tracks
- Access for emergency vehicles
- Open space availability
- Favorable topography
- Resources
- Ample roads
- Power
- Telephones
- Shelter

- Water
- Visibility—line of sight to exclusion and CRZ/C zones
- Prevailing wind direction up-wind of the exclusion zone
- Distance, as far as practical from the exclusion zone
- Type of work being performed

5.7 WORKER COMFORT AREAS

Worker comfort areas can be located in site work zones. These comfort zones allow workers to take breaks and rest without being contaminated. These areas are designed to maintain the safety of workers and generally require special procedures for ingress and egress, personnel and air monitoring, potable water consumption, and restroom use [1].

5.8 LESSONS LEARNED

The names, number, and types of zones vary based on the activities at the worksite. The important thing to remember is that work zones are established to protect the workers and the public. Everyone on the site should understand the hazard(s) and control(s) necessary to support each identified zone. Wind direction was mentioned as an important criterion in choosing the support or clean zone. In most parts of the country, wind direction can be highly variable. If this is the case at a given site, how should the support zone be handled? The answer may vary based on the conditions.

To illustrate this point, let's consider the example of a superfund chemical waste landfill remediation job. In this example, we will need to determine the contaminants of concern. After making this determination, we next need to assess the contaminants and how they may migrate from the site. We need to determine if changes in the wind direction occur, how frequently, and how workers in the support zone will be affected. In most cases, the likelihood for workers in the support zone to be affected is minimal.

However, if the cap on the landfill contains a hazardous material such as lead containing dust that is being transported off site, there are a couple of questions to ask. The first is, "What type of work activity will be taking place on site?" And second, "Does work activity create dust?"

If dust is being created, alternatives should be considered. One alternative may be a change of location of the support zone. This may be more difficult than it sounds. Many times the support zones are trailers, buildings, or structures that are not portable, or at least are hard to move from location to location.

Another alternative might be to make sure that dust suppression, such as water or foam, or other controls are instituted so that the wind will not transport hazardous materials to the support zone. This is easier said than done. Although dust suppression techniques have been used with success, if there is a "glitch" in the dust suppression system, workers in the support zone may be exposed. It appears that this situation might be more difficult to resolve than initially thought. This theoretical problem has existed on many hazardous waste sites. We believe that this situation could have best been resolved during the planning stages.

For most hazardous waste sites with proper planning the situation is known before remediation activities begin. The support zone location needs to be carefully considered at the planning stages of the project. A better solution to the theoretical problem at hand would be to move the support zone further from the source of contamination. If the support zone can be placed far enough away, the likelihood for exposure at the support zone is minimized.

However, moving the support zone farther from the source of contamination often brings up logistical problems associated with the distance. The logistics should be considered at all times. No matter how open the space is, there are always distance constraints.

Let's look at another example: a small-scale voluntary cleanup that might take place in the lot behind a factory, or a tank being removed at a corner gas station. Although we have the same considerations, these projects are on a smaller scale and will create less of a hazardous condition. The same principles that exist at the large job still should be adhered to on the small job. The work zone should be delineated and controlled to protect workers or the public from entering the work zone. For the small job, caution tape, snow fence, or traffic cones can be used effectively.

Personnel and equipment need to be decontaminated in the CRZ. However, the CRZ might be a small area immediately adjacent to the remediation area, which workers are aware of, and is also marked appropriately. Although the CRZ is less formal and likely does not have many decontamination stations, efforts should be made to make sure that personnel and equipment are appropriately cleaned. Many times, due to the logistics of a smaller job, disposal of wastes becomes difficult. If purge water is drummed and left on the site, it is imperative that it is identified, labeled properly, recorded in the site log, and disposed of in the proper manner (in accordance with applicable, local, state, federal, or other regulations).

As should be clear now, determining work zones can be a challenge. You can run into unique situations performing hazardous waste remediation work. Let's consider another landfill site that is bordered on one side by a river and on the other side by a railroad track. Although choosing the support zone location is not a problem, determining how an injured worker might be transported from this site can be a challenge. We previously mentioned that in choosing the support zone, we should ensure close proximity to highways and railroad tracks, easy access for emergency vehicles, sufficient available open space, and favorable topography. This should be one of your primary concerns. If workers are expected to work near the river, you should provide a life-saving skiff and two ring buoys with 90 feet of rope. All workers should wear Coast Guard–approved life preservers.

We previously discussed space and topography as a factor. For our current site situation discussion, after we met with our local emergency planning committee and local law enforcement and hazardous material responders, we could choose to build a helicopter landing pad so that any potentially injured workers could be transported to a medical facility for treatment if the need arose. The site management could also obtain a radio so that they could have immediate contact with the train dispatcher. The dispatcher could have a crew uncouple railcars and move them so that heavy equipment could be brought to the site.

We earlier discussed ample roads, power, telephones, shelter, and water. For remote sites, roads can be built, portable generators can provide a tremendous amount of power, cellular telephones can provide communication almost anywhere, and shelter and water usually can be easily obtained. We also discussed visibility, and line of sight to exclusion and CRZ/C zones. Keep in mind that line of sight these days, like the buddy system, should not be taken literally. Site radios, cell phones, or both, if properly used, can assist with the buddy system and line of site. Site radios usually work out well for the observer to obtain assistance, if needed. In addition, the cost is typically substantially less than cell phones. However, cell phones have a distinct advantage that is priceless—you can immediately contact outside emergency services in case of a site emergency. Should it be radios, cell phones, or the buddy system? This is an important point which should be determined in the planning stages. It cannot be stressed enough that planning is the key to any successful project.

5.9 TRAINING

Training plays a huge role in ensuring that site work is performed safely. Training is even more important when workers are dealing with or may be exposed to hazardous materials. Training becomes more complicated in that case because of HAZWOPER and other regulatory guidelines.

Due to the importance of training as it relates to hazardous waste, all of Chapter 8 is dedicated to this subject.

5.10 DETERMINING APPLICABILITY OF OTHER REGULATIONS AND REQUIREMENTS

In addition to the hazardous waste standards, there may be a variety of other standards that may apply to any worksite. One standard that seems to surprise PMs is the lead standard. Even projects that are "clean construction" (not HAZWOPER or working with chemicals) may come under the lead standard. If new process equipment is being installed in an existing plant, any tie-ins, pipe rack, structural members, and even walls may have been covered with many layers of lead-based paint.

As an example, if the work requires that the lead-based paint is to be disturbed by drilling a hole in a beam (the beam that has been painted with lead-based paint), will workers be exposed to lead in the air? What should be done under the standard to be sure that workers are not being overexposed? Under the standards 29 CFR 1910.1025, "Occupational Exposure to Lead in General Industry," and 29 CFR 1926.62, "Occupational Exposure to Lead in Construction," the following are some criteria that should be applied to determine who should be enrolled in the lead program.

Construction jobs are often of short duration, and, without sufficient protection, workers could be exposed to high concentrations of airborne lead during the period between sampling and receipt of the results. For these reasons, OSHA requires that the decision to enroll a worker in a special medical program addressing potential lead exposure depends on whether the worker is engaged in an OSHA-listed task, not on measured airborne exposure levels [4]. OSHA has established a hierarchy of three lists of tasks, the performance of which, in the presence of lead, trigger basic protective provisions before airborne lead monitoring. All three sets of tasks mandate initial medical surveillance consisting of blood sampling and analysis.

The General Industry Lead Standard imposes medical program requirements when an employee has the potential to be exposed above an action level for more than 30 days. OSHA's three sets of tasks differ mainly in the level of respiratory protection required for workers occupationally exposed to lead [5].

Workers engaged in any of the listed tasks require initial medical surveillance consisting of blood sampling and analysis. Protective measures, including graduated levels of respiratory protection and PPE tied to the task grouping, change areas, hand-washing facilities, and training, should be provided to workers performing any of the tasks. It is not necessary to collect new monitoring data each time because OSHA's analysis of previously collected exposure data already indicates that high exposure levels may be expected when these tasks are performed. Biological samples that are collected should be analyzed by an OSHA-approved laboratory, and results should have an accuracy of +/−15 percent or +/−6 micrograms per deciliter (g/dl) blood [5].

What happens if biological monitoring results exceed the benchmark? Medical removal and medical removal benefits should be provided under certain conditions. The General Industry Lead Standard and the Construction Industry Lead Standard contain slightly different provisions requiring the medical removal of an overexposed employee. The General Industry Lead Standard requires removal based on the average results of three blood tests in excess of 50 g/dl. The Construction Industry Lead Standard, however, stipulates two triggers for medical removal. Medical removal is indicated if the employee is exposed at or above the airborne action level and in the event of either of the following: (1) if a periodic and follow-up blood-sampling test equals or exceeds 50 g/dl, or (2) if a medical finding or opinion documents that the employee has a detected medical condition placing the employee's health at increased risk from exposure to lead.

If a worker is overexposed to lead and should be removed from the job due to exposure, the standard specifies medical removal benefits and more [5]. If a worker claims to have been overexposed to lead, will you be prepared to defend that claim? If a worker notifies regulators that he or she has been exposed to lead, will your program stand up to scrutiny?

Besides lead, there may be a variety of other substances that a worker may be exposed to. Earlier in this chapter we mentioned the possibility of mobile laboratories. If your site does have a mobile lab, there are a variety of other regulations that should be considered. Of course, hazard communication comes into play. A typical laboratory can have volumes of material safety data sheets in its libraries. Besides hazard communication, you should determine whether a chemical hygiene plan is a requirement. In addition, we need to consider how the laboratory might affect the site emergency plan.

The types of programs needed to protect workers should be determined far before the work begins. Preparation is again the key. If we know that we are going to sample for lead, our HASP will specify how, where, when, how often, and so on. If equipment is needed to institute these programs, it should be considered during the planning phases. If special talents are required to perform sampling and other tasks, the qualifications of these persons should be determined prior to HASP development. If we know we are going to have a lab on site, we should consider the effects on the HASP and other regulatory requirements. A properly researched, comprehensive, well-written HASP will provide for greater worker protection and minimize later surprises.

REFERENCES

1. *Handbook for Occupational Health and Safety During Hazardous Waste Activities.* Office of Environmental, Safety and Health Office of Environmental Management, 1996, pp. 3-11, 3-13, 6-1, 6-5, 7-7–7-9.
2. *Working Safely During DOE Hazardous Waste Activities.* U.S. Department of Energy, June 1996, p. 11.
3. 29 CFR 1910.195 "Occupational Exposure to Noise."
4. 29 CFR 1910.62 "Occupational Exposure to Lead in Construction."
5. 29 CFR 1910.1025 "Occupational Exposure to Lead in General Industry."

Chapter 6

Development of a Site-Specific Health and Safety Plan

When the appropriate research has been completed, it is time to use the information to develop the site-specific safety plan. Keep in mind that this plan will provide the basis for protection of workers, visitors, and the public. The plan defines health and safety hazards, controls, and requirements for individual activities at a specific worksite and provides documentation to help identify and control health and safety hazards before fieldwork begins [1].

6.1 LENGTH

A HASP should not be a health and safety program (as discussed earlier) nor should it be a lengthy, all-encompassing document. Experience has shown that HASPs vary from nonexistent, to terse, to encyclopedia-length documents. Although not the rule, typically the larger, more complicated, and more hazardous a site, the more extensive the HASP [2]. Another important factor to remember when determining the length of the HASP is the development of the safety culture. Management in a poorly developed safety culture may believe that HASPs are not necessary. In this type of safety culture the HASP length will typically range between none and terse. In fact, you might be asked the following question: "Why should we spend time, resources, effort, and money on a document that we do not need?" Let's examine this question as it relates to HASP length.

Most requests for proposals and bid specifications will include statements that all work will comply with all applicable safety guidelines. You may come across the argument that, "Since we have already agreed to abide by the law (OSHA standard), why not just submit the latest copy of 29 CFR or another applicable guidance and include it as an attachment to our work plan, and save the time and effort of developing a safety plan?"

We believe (and regulators agree) that attaching 29 CFR or other documents as a substitute for a safety plan is not compliant, nor is it a good way to promote safe work activities. We also believe that the HASP length and complexity should consider the work activity, duration of activity, and hazard on the site. We will discuss this point in more detail later in this chapter.

6.2 SPECIFIC HASP WORDING

The HASP should apply to the site-specific work activities. To be most effective, the HASP should be prepared in concise, to-the-point terms. The object is to make the HASP as simple as possible, so that everyone can understand the contents. Language requiring interpretation should be avoided.

The object is to include sufficient details of the work area being referred to. For each work area, specify types of PPE required or levels of protection that will be required when doing the tasks in those particular areas.

Keep in mind that you should be familiar with the type of work that you will be doing before you do it. However, try as you might, the unexpected can and often does occur. Therefore, as soon as the unexpected occurs, you must react. The reaction should include a hazard assessment of the unexpected work activity. One effective way to do this is through a job hazard analysis. Note: JHAs were discussed in detail in Chapter 4.

The HASP builds on and enhances existing health and safety program components. In describing PPE, generic descriptions of Levels A, B, C, and D should be avoided. Instead, define each level for the specific area or activity in question. Typical questions concerning HASP development are summarized in the following sections [1].

6.3 ELEMENTS

A properly written safety plan contains worker health and safety program information, guidance, and alternatives. The HASP should quickly answer the two questions: "What hazards are present?" and "What provisions have been made to make sure that all tasks will be performed safely?" [1] Subsequent chapters will provide a detailed examination of a typical HASP. The information presented will be generic and should be modified to fit any site-specific hazards.

Hazardous waste operations often include tasks and activities that are conducted on a periodic basis, are of very short duration and transient in nature, or otherwise pose little hazard. Developing a brief HASP template (e.g., "fill in the blank"), a permit, or a checklist system that includes essential HAZWOPER-type information may suffice for these types of operations. A HASP requires certain basic information as mandated by existing HAZWOPER, DOE, and Army Corps of Engineers. To get the maximum benefit from these requirements, there are specific elements that should be incorporated in a HASP. The following sections will highlight and suggest some information that can be presented that will comply with specific requirements.

6.3.1 Cover Sheets

Although a cover sheet is not mandatory, it is recommended and can be effective to make sure that some type of sign-off is incorporated. This following information should be included:

- The name and location of the site
- What entity is authorizing the work
- The name of the author of the HASP
- The date of HASP finalization
- Approval(s)

Approvals are an important part of the cover sheet. For the large site, there should be a minimum of five levels of approvals for each HASP. The most important include:

- PM or project director
- Site supervisor
- Contractor/subcontractor
- The health and safety professional who authored the HASP
- The client

Other signatures may also be required. For example, if there are subcontractors who will be performing site work, a representative from the subcontracting firm should review and approve the plan. If the site has oversight contractors present, is complicated, large in size, or includes work that is projected to take place over a long period of time, there will likely be additional approvals required.

Keep in mind that the number of signatures means little compared to the content of the plan and how it is executed. A HASP may be of excellent quality, but if execution is poor, then the likelihood for mishap increases. On the other hand, obtaining approval signatures alludes to buy-in, understanding, and agreement. Your chances of approval signatures meaning buy-in, understanding, and agreement increase significantly if the HASP is distributed to those who are to approve it at least two weeks prior to approval requirement. For larger sites, there may be a 30-day or more notice for the approval requirement.

An indicator that someone is truly interested in the HASP prior to approval is in the comments or questions received prior to placement of any signature.

Although signatures are no guarantee that you will do the job any more safely, not obtaining signatures is not an acceptable alternative. We believe in signatures and feel that they should be a requirement for safety plans as well as many other safety-related documents such as:

- A statement of understanding and compliance for workers completing site orientation
- Daily safety meetings
- Training sessions
- Safety inspection and safety violation documents

Signatures should be obtained and retained on file. Although an unlikely occurrence, you may rarely find a worker who refuses to sign a document. This signature refusal experience can be traumatic, especially for the inexperienced manager. We recommend that before you get upset, you try both adjourning to a neutral corner and ironing out difficulties. You may find it helpful to get a few more people involved so that a lively discussion ensues.

You should try to determine the true reason why someone has refused to sign the document. You may find that the worker who refused to sign has one or several valid points of contention. You may consider adjusting your program or presentation to address these points. Alternatively, you may also disagree with the reasons offered for refusal to sign. This situation needs to be brought to the attention of management, human resources, or others, depending on the organizational structure. You may find that your immediate best alternative is to note that the worker refused to sign, and continue with site work. This situation may, however, have far-reaching implications and legal ramifications. Getting help and giving quick notice to the right people in your organization should be considered. We offer no further thoughts on this situation, and hope that the problem never befalls any of you.

6.3.2 Introduction

The introduction is a brief statement regarding the development of the HASP. It should include the applicability and limitations. In this section, a statement is typically made that sets the stage for the safety plan and disallows any changes to the document without an amendment being completed and approved.

6.3.3 Site Description/Background Information

This is important information that describes the site and provides workers, visitors, and other personnel with pertinent site information. In addition to studying job specifications, contracts, and talking with project management, the author(s) should develop a detailed operating history of the site. The history is useful when determining potential site hazards. The type of information that can typically be located includes:

- Types of wastes that were accepted
- Years of operation
- Any operating permits
- Ownership of the property and previous owners
- Complaints or regulatory background
- Results of previous studies
- Results of analytical information
- Known or suspected hazards present
- Surrounding topography
- Surrounding community involvement
- Other items of concern

If the facility operated as a manufacturing plant or other entity, you might consider including the following information:

- Products manufactured
- Years of operation
- Ownership, previous owners
- Operating or other permits held
- Surrounding topography
- Title search information

If you have conducted adequate research prior to authoring the plan, this section will not provide an insurmountable challenge. However, if the required information is not readily available or not understood, the information should reflect what you have found.

Each worksite is different even though relative hazards may be very similar. Wind direction, surface contamination or neighboring properties that may contain contributing contamination vary [3]. The site description/background section should provide workers with enough information to perform their job safely without overkill. The simpler, smaller, and fewer hazards there are on site, the less background information will be necessary.

6.3.4 Project Personnel and Responsibilities

HAZWOPER specifically requires that project personnel and responsibilities be well defined. Refer to our discussion on project team organization in Chapter 3.

6.3.4.1 Site-Specific Hazard Analysis

Each hazard is analyzed and documented as specifically as possible in this section. Specific job tasks and hazards associated with those tasks should also be included. If analytical information is available for site contaminants, it should be included. These typical hazards may also include physical, chemical, biological, and radiological, as discussed in the next sections.

6.3.4.1.1 Physical Hazards

You should always anticipate hazards such as sharp objects like nails, broken glass, and medical needles; slippery surfaces; steep grades; and potentially unstable surfaces such as walls, floors, or roofs that could cause falls, give way, or collapse. Other common physical hazards include:

- Material handling
- Operating machines and heavy equipment
- Excavations (holes and ditches)
- Electrical sources
- Confined spaces
- Fire and explosions
- Heat and cold stress
- Noise
- Poorly stacked or unstable drums [2]

In most cases, physical hazards are easy to identify. Let's consider housekeeping items. These items can contribute to slip, trip, and fall hazards. Most people are comfortable in observing and fixing these types of hazards, especially after an accident occurs. After all, you can easily see these types of hazards. The accident occurrence is also easy to envision.

If our inspection process has pointed out areas in need of housekeeping, and these same areas continually "show up" on our inspection, an adjustment would appear to be in order. One possible solution would be to spend more time and effort on housekeeping issues. However, no matter how much time and effort we spend on housekeeping, we can usually find places lacking in housekeeping. Possibly there is no money, time, equipment, or other resources available to perform housekeeping activities at this time.

Consider some alternatives. We should consider using control access zones (CAZ) to limit worker exposure to zones where slip, trip, and fall hazards exist. By limiting worker exposure we should be limiting accident occurrences. If equipment is a temporary problem, we should consider leasing or rental. If labor is a problem, we might consider utilization of temporary employment agencies.

6.3.4.1.2 Chemical Hazards

Handling hazardous chemicals has become part of most people's everyday living. Just consider gasoline, and how most people fill their own tanks. In the manufacturing arena, chemicals are commonplace. On hazardous waste sites there are a variety of unknown chemical substances and other hazards that may take the form of a solid, liquid, or gas. The effects of exposure to toxic chemicals may either be immediate (e.g., acid burns) or delayed (e.g., lung damage from inhaling asbestos). There are four routes of chemical exposure that exist:

- **Inhalation:** Most common means of entry.
- **Skin or membrane absorption:** Chemicals can be absorbed through intact skin or the eyes.
- **Unintentional injection:** Chemicals can enter the body through open wounds or accidental punctures.
- **Ingestion:** Chemicals can be ingested on the worksite by eating, drinking, or smoking.

Other specific chemical hazards that workers may come into contact with are too numerous to mention. The effects from these chemicals vary widely. It is important to know if there are chemicals being brought on site for any reason, along with the chemicals already at the site and chemical wastes present on site. After you have obtained a comprehensive chemical library, you should determine compatibility and synergistic, additive, and other effects of chemical mixing. This might include fire, explosion, or release.

6.3.4.1.3 Biological Hazards

Biological hazards can result from exposure to insects, animals, plants, bacteria, and various viruses. Any particular site may include a variety of biological hazards such as:

- Bites and stings from spiders, insects, snakes, and other wildlife
- Skin rashes and allergic reactions from contact with poisonous plants or animals
- Infections from contact with or exposure to bloodborne pathogens or other biological agents in contaminated soil, waste, dust, bird and animal droppings or transmitted by insect bites or stings

6.3.4.2 Assessment Hazard Identification and Control

Critical to hazard characterization is the identification of hazards and the assessment of possible worker exposure. This can be accomplished in a variety of ways. As described before, one commonly used technique is a JHA with project teams that include the worker. The information collected is used by the SSHO and the radiation control officer to develop an appropriate hazard control and protection strategy.

NOTE: We dedicated Chapter 4 to JHAs. Although government literature refers to JTHAs, we believe that, in principle, they are equivalent. We will be using JHA instead of JTHA or other terms.

The HASP should contain the information obtained during the preparation phase concerning hazard characterization and exposure potential. If the information has gaps, ranges, or is incomplete, this should be taken into consideration so that proper protective measures are taken. If and when new information is discovered, this should be included as part of the hazard characterization as an amendment.

Hazard controls include engineering and administrative controls and PPE. Hazard characterization is a tool that is used to develop hazard controls and safe work practices and procedures and to make sure that the appropriate PPE is selected for each job.

The HASP should describe how potential health and safety hazards at the work site are located, identified, and measured. A written schedule including inspections and walk-throughs conducted by designated individuals should be specified.

JHAs of individual work operations or tasks and their associated hazards should be included in the plan to help develop effective controls. Many times subcontractor activities are added to the safety plan after its original publication. Typically, subcontractor activity is specialized and short lived. It would be advantageous to have all JHAs completed and included as part of the work before the work begins. For HAZWOPER work, it is mandatory that subcontractor activities be covered in the HASP prior to work inception. For late arrivals of JHAs, a HASP amendment should be initiated and redistributed to all parties [1].

Each worksite may use various kinds of monitoring instrumentation to identify and measure levels of different types of hazards that may be present. These are discussed in greater detail in the next section.

After potential hazards have been identified, site access and hazard controls should be developed and put in place before work begins. This process of recognizing and evaluating new hazards and putting controls in place continues until the task or job analysis is complete. In addition, as mentioned earlier, as new information is discovered or becomes available, this should be immediately considered. If an amendment is in order, this should be completed and communicated to all involved.

6.3.4.2.1 Engineering Controls

Engineering controls are designed to eliminate or keep hazards away from a person. Examples include machine guards on equipment, ground fault circuit interrupters, local exhaust ventilation that removes contaminated air at the source, and remote systems (like robotics) used to handle dangerous materials.

6.3.4.2.2 Administrative Controls

Administrative controls include limiting the time spent in a hazardous area, SOPs, proper designation and posting of areas, or changes to work practices. Other examples include identifying and limiting entry into confined spaces and using lockout/tagout procedures.

6.3.4.3 Exposure Monitoring

Exposure monitoring should be developed based on site-specific information as a result of all the information gained from the preparation phase. We cannot overemphasize the importance of using only qualified individuals to develop exposure monitoring strategies. In addition, only trained and qualified field personnel should operate screening equipment and be allowed to interpret results [3].

Whatever type of monitoring instrumentation is employed it should be operated, calibrated, and maintained in accordance with all recommended manufacturer specifications. A copy of the operating manual should be maintained in close proximity to the equipment and should be included as an appendix to the safety plan. Those who are operating the equipment should be trained adequately and understand the limitations of that equipment. The operator should know the contents of the manufacturer's manual and be able to answer questions about that equipment.

6.3.4.4 Chemical Hazard Control

Some general control procedures are offered in Chapter 4. The handling procedures adopted should ensure that whatever the hazards present, they are handled adequately. Because this list is general in nature, the user will have to determine applicable site-specific control procedures.

6.3.4.5 Hazard Communication

Site-specific information pertinent to hazard communication should be included in the HASP. For instance, if there is concern over metal contamination in site soils and dust, this information should be clearly disseminated to anyone who may come into contact with it [4]. Even for those persons who will likely not come into contact with it because of administrative controls, some training regarding access zones might be in order. For instance, when traveling on site, only use roads marked for general use. Not doing so might put you in an area with a potential for contaminated metallic dust exposure.

6.3.4.6 Personal Protective Equipment

Refer to Chapter 9 for a detailed discussion on PPE.

6.3.5 Site Control/Work Zones

Hazardous waste sites are divided into as many or as few zones as necessary to protect worker health and safety. Work zones are established to prevent the spread of hazardous substances from contaminated to clean areas. Radiological work zones should be considered compatible with hazardous waste work zones, differing only in terminology. Diagrams, sketches, and maps should be used as often as necessary and constantly updated and communicated so that workers can be sure that they are appropriately protected [3].

Work zones are designed to control access to actual and anticipated hazards. Work zone positioning is based on hazard characterization and exposure assessment. Anticipated work activity, potential releases, and the amount of contaminant dispersion are important for delineating these zones [3].

6.3.6 Buddy System

No one should enter a contaminated area or an exclusion zone without a buddy (someone who can aid you in case of an emergency) who is capable of the following:

- Providing the partner with assistance
- Observing the partner for signs of adverse exposure to chemical, physical, or radiological hazards
- Notifying the appropriate persons if emergency help is needed
- Periodically checking the integrity of safety systems and the partner's PPE and other safety equipment [3]

6.3.7 Decontamination Procedures

Decontamination is a process that is site specific. Meteorological conditions may, at times, have an effect on the decontamination process. Rainy conditions may produce mud. The mud not only makes the work more challenging, but also the decontamination process, since mud is typically mixed with waste.

A tremendous amount of work has to be done to make sure that effective decontamination is accomplished. However, the site decontamination process should be constantly reviewed to make sure of its effectiveness. This process should be continuous.

We do not attempt to discuss in depth decontamination methods for radiological wastes. A health physicist should be immediately available to assist with decontamination of radioactively contaminated personnel or equipment.

Our discussion here is for the typical petroleum-based waste or low hazard chemical waste. For this situation we prefer the common-sense approach to the handling of hazardous materials. Whatever process is effective in making sure that the hazardous materials stay on the site and are disposed of in an appropriate manner should be utilized.

The typical decontamination may include removing any gross contamination in the exclusion zone using equipment that will stay in the EZ (for instance, a hand scraper, a wire brush, etc.). Once gross contamination is removed, the worker (or equipment) might go to the "decon pad" where washing with a scrub brush, soap, and water might take place. Chapter 10 provides an in-depth discussion of decontamination and work zones.

Equipment may get washed with a steam jenny and allowed to air dry on plastic in a more remote area. We need to keep in mind that steam cleaners have the potential to cause substantial physical harm. The combination of high-pressure water and high temperatures can be dangerous. When this is coupled with a worker standing on visquine or plastic, it becomes a slip, trip, and fall situation. Situations compounded with respiratory and whole-body protection, such as saranex or rubber suits and gloves, add in the potential for poor vision, heat stress, and the lack of physical dexterity. Keep these issues in mind prior to steam cleaner activity.

Disposable PPE should be removed, and the workers should thoroughly wash and rinse themselves. Anything contaminated should be left on site and disposed of in the proper manner. In this case, the worker and equipment would leave the site only after having been thoroughly cleaned. Refer to Chapter 10 for detailed decontamination activities.

6.3.8 Training

Training requirements should be addressed in the site-specific HASP. For larger, more complicated sites, training matrices may be used so that different levels of training can be appropriate for different phases of work activity. Refer to Chapter 8.

6.3.9 Medical Surveillance

Medical surveillance programs are designed to accomplish the following goals:

- Demonstrate that workers are fit to perform their jobs safely and reliably
- Provide ongoing assurance that access and hazard controls limit worker exposure
- Comply with requirements set forth by OSHA, DOE, the Army Corps of Engineers, or other agencies

A medical surveillance program is designed to protect the workers' health. Given the limitations of industrial hygiene monitoring data and the many hazards involved in hazardous waste activities, medical surveillance data may provide the only indication that worker exposure to toxic substances has occurred.

Medical monitoring and surveillance programs enable occupational health professionals to identify adverse health effects caused by exposure to hazardous substances and conditions and to discuss plans with site workers, industrial hygienists, safety professionals, and line management to prevent exposures and protect workers. These goals can be accomplished through two objectives:

- Detection of preexisting diseases or medical conditions that place employees performing certain tasks at increased risk
- Control of individual workplace exposures in a manner that minimizes adverse health effects [3]

Although OSHA, DOE, and the Army Corps of Engineers establish the elements of a medical surveillance program, the occupational health physician is responsible for determining the content of medical surveillance examinations [5]. The health and safety staff is responsible for providing all exposure monitoring data and other technical support needed by the physician to implement the program properly, and any radiological control organization is responsible for providing worker external and internal radiation exposure measurements and other technical support that may be necessary.

Medical surveillance programs range from support contracts with local hospitals or physicians to full-scale on-site occupational health organizations that include physicians, nurses, and technicians who are employed by prime contractors. The option selected depends on the size of the project, the nature of the hazards involved, the capabilities of local facilities, and the resources available.

Regardless of the option selected, worker occupational health records should be provided to the site's occupational health physician, thereby facilitating the availability of, and access to, adequate medical care in the event of an emergency. Provisions that are consistent with current regulations pertaining to privacy should be made to retain these records after completion of project activities. OSHA regulations mandate that, unless a specific occupational safety and health standard provides a different time period, the employer should meet the following criteria:

- Maintain and preserve medical records on exposed workers for 30 years after termination of employment.
- Make available to workers, their authorized representatives, and authorized OSHA representatives the results of medical testing and full medical records and analyses.
- Maintain records of occupational injuries and illnesses and post an annual summary report.

General guidance for designing medical surveillance programs can be found in the HAZWOPER standard and medical surveillance requirements for several specific substances as provided in 29 CFR Part 1910, "Occupational Safety and Health Standards," Subpart Z, "Toxic and Hazardous Substances." Whenever multiple standards affect worker health and safety, the more protective requirements should be followed. These determinations should be made by knowledgeable health and safety professionals. Occupational health physicians providing medical surveillance support for HAZWOPER sites are to be provided with copies of the HAZWOPER standard.

An outline of the medical surveillance program, as approved by the occupational health staff, should be incorporated in, or appended to, the site-specific safety plan. Modifications to the program should be based on the professional judgment of the occupational health physician, in consultation with the health and safety professionals, and on the hazards of the specific worksite.

Changing working conditions that require modifications to medical surveillance activities can be communicated to the medical department by a supervisor through the health and safety organization and the personnel department, where records are maintained. This should include regular visits to worksites and facilities by occupational medical physicians and selected medical staff to familiarize themselves with tasks and actual or potential hazards. Contractor management should require participation by medical personnel in new materials and process review committees, safety committees, and other health-related meetings. The medical surveillance program should be reviewed regularly to make sure that it is effective.

The SSHO should on an annual basis in cooperation with the occupational medical physician and the health and safety professional conduct the following:

- Ascertain that each accident or illness was investigated promptly to determine the cause and make necessary changes in health and safety procedures.
- Evaluate specific medical testing to determine potential site exposures.
- Add or eliminate medical tests as indicated by current industrial hygiene and environmental data.
- Review potential exposures and the HASP to determine if additional testing is required.
- Review emergency treatment procedures and update the list of emergency contacts.
- Ensure timely employee access to records on their request.

Existing respiratory protection or hearing conservation programs can be referenced and integrated, as appropriate, into the site-specific medical surveillance program after worksite hazards have been considered. At some sites, workers are provided a fitness-for-duty card indicating their current medical status and the medical surveillance programs in which they participate [1].

6.3.9.1 Workers Included in Medical Surveillance Programs

HAZWOPER, related DOE, and the Army Corps of Engineers rules and requirements stipulate that employees involved in any of the following activities who have a reasonable possibility of exposure to hazardous substances or health hazards at specified levels (see 1910.120 [f][2]) should be included in a medical surveillance program:

- Voluntary cleanup operations, or those required by DOE or the Resource Conservation and Recovery Act (RCRA), or as otherwise defined by the HAZWOPER Standard
- Treatment, storage, and disposal (TSD) operations, as defined by the HAZWOPER Standard
- Operations at hazardous waste activities worksites where use of a respirator due to potential radiological (as specified by Article 532 of the Draft DOE Radiological Control Technical Standard) or nonradiological exposure is recommended or required
- Operations resulting in potential exposure to a regulated chemical or radiological agent, as prescribed by DOE and OSHA action levels, or to a bloodborne pathogen
- Operations requiring use of a respirator for 30 days or more per year or resulting in an exposure that may be at or above an OSHA PEL,

or if there is no PEL, above the published exposure levels (whether or not a respirator is worn) (see an exception discussed in 29 CFR 1910.120[f][2]).
- Hazardous waste or emergency response activities resulting in injury, illness, or signs or symptoms of possible overexposure to hazardous substances or health hazards from those activities

The following employees should also be included in a medical surveillance program:

- Individuals who respond to emergencies involving hazardous wastes, including hazardous materials (HAZMAT) team members
- Any employee who exhibits signs or symptoms that may be the result of exposure to a hazardous substance [1]

6.3.9.2 Frequency and Content of Medical Examinations

Before work activity begins, all employees required to participate in a medical surveillance program for hazardous waste activities should undergo a baseline medical examination (a physical exam). This exam should be based on specific hazards identified during the preliminary hazard assessment. Periodic follow-up exams are required at the discretion of the attending physician. Typically, these follow-ups are completed annually, however, they can be adjusted to more often or less often dependent on the exposure [1].

Based on the professional judgment of the occupational health professional, more frequent examinations may be required. An examination may be required when a worker changes jobs or tasks. To facilitate this process, a representative of the medical staff should be invited to attend management and worker briefings or meetings and should participate as a member of a worker protection team. For small sites or small companies, there will be no worker protection team. However, whoever is in charge of safety and health should invite the occupational physician to the site so that they get a feel for potential exposures.

Medical surveillance may need to address much more than the basic requirements in the HAZWOPER standard. Based on the presence of hazards (such as lead, asbestos, and carcinogens), special types of medical testing may be required. The occupational health physician responsible for the medical surveillance program should work with the rest of the medical surveillance team to determine what forms of surveillance are applicable for activities at each worksite.

Medical examinations and consultations should be provided to the employee without cost, without loss of pay, and at a reasonable time and place. The content of the examination or consultation is determined by the occupational health professional, based on information provided by the health and safety staff. Employees performing on-site hazardous waste operations or entering an exclusion zone or contamination reduction zone at a hazardous waste site are required to receive specific medical examinations at designated intervals.

For activities beyond those explicitly addressed by HAZWOPER and for activities for which more than one regulation is relevant, use the regulation that is more protective of worker health and safety. These provisions should be incorporated into the medical surveillance program [1].

6.3.10 Emergency Treatment

Both emergency and acute, nonemergency medical treatment should be available at the worksite. The plan should be integrated with the overall site plan and the surrounding community emergency and disaster plan. In addition, input from and review by the occupational medicine physician and health and safety personnel is invaluable for developing the medical and emergency preparedness portions of the plan.

The plan should include a list of all potential hazards and their locations, personnel responsibilities, and actions to be taken in the event of an emergency. Emergency medical treatment should be integrated into the overall site emergency response program. Individual worksite managers should contact the site emergency preparedness group to verify that all potential emergency responders and care providers understand the hazards of the worksite and can be relied on to provide services as needed.

The following guidelines for establishing an emergency treatment program should be documented or referenced in the safety plan:

- Train a team of site personnel in emergency first aid.
- Train personnel in emergency decontamination procedures in coordination with the emergency response plan.
- Designate roles and responsibilities.
- Establish an emergency/first-aid station on site.
- Arrange for a 24-hour on-call physician.
- Establish an on-call team of medical specialists for emergency consultations.
- Develop a protocol for handling thermal stress and other potential health disorders.
- Make plans in advance for emergency transportation to and treatment at a nearby medical facility.
- Post names, numbers, addresses, and procedures for contacting on-call physicians and medical specialists.
- List ambulance services, medical facilities, poison control, and fire and police services.

- Provide maps and directions to the nearest medical facility.
- Establish a radio communication system for emergency use.
- Review emergency procedures daily with all site personnel at safety meetings before beginning the work shift.

Nonemergency medical care should be arranged for hazardous waste site personnel who are experiencing health effects resulting from an exposure to hazardous substances. Off-site medical care should make sure that any potential job-related symptoms or illnesses are evaluated in the context of the employee's exposure. Off-site medical personnel should investigate and treat non-job-related illnesses that may put the employee at risk because of task requirements [1].

REFERENCES

1. *Handbook for Occupational Health and Safety During Hazardous Waste Activities.* Office of Environmental, Safety and Health Office of Environmental Management, 1996, pp. 3-9, 6-1–6-9; 9-2, 9-3, 9-5, 9-7.
2. *Hazards Ahead: Managing Cleanup Worker Health and Safety at the Nuclear Weapons Complex.* U.S. Congress Office of Technology Assessment. Washington, DC: U.S. Government Printing Office, 1993, pp. 3, 66.
3. *Occupational Safety and Health Guidance Manual for Hazardous Waste Site Activities.* Prepared by National Institute for Occupational Safety and Health (NIOSH), Occupational Safety and Health Administration (OSHA), U.S. Coast Guard (USCG), U.S. Environmental Protection Agency (EPA), October 1985, pp. 5-5, 6-1, 6-2, 7-1, 9-3, 9-4.
4. 29 CFR 1910.1200 "OSHA Hazard Communication Standard."
5. 29 CFR 1910.120 "OSHA HAZWOPER Standard."

Chapter 7

Implementing the Safety Plan

After the safety plan has been completed and approved by the management team, the most challenging part of the job needs to be addressed. It is, very simply, the execution process. Now that the plan is written we should make sure that all site work is performed in a safe manner. Worksite controls established in the plan should come into play immediately when activities begin. It is essential that everyone at the worksite is aware of the contents of the safety plan. To make sure that everyone is familiar with the safety plan contents, everyone should be oriented before any work is performed.

To make sure that safety is a priority at your project, the safety plan needs to be adhered to. All workers should become familiar with and trained in at least those parts of the safety plan that may affect them. Workers should not be deemed qualified to perform their assigned job functions until site management is satisfied that they have received not only the required functional training, but other safety-related site-specific instructions.

7.1 ORIENTATION

An effective orientation is the first step in making sure that workers understand what is expected to perform their work as specified in the plan. The details of the orientation should be worked out during the pre-planning session.

The site orientation program will set the tone for your project. An organized, well-thought-out, and comprehensive orientation will get workers off to the best start. On the other hand, if the orientation is weak, haphazard, and poorly directed, this will be a reflection of the organization in charge and will likely be difficult to overcome.

A more complicated or dangerous site will require a more extensive orientation. A seasoned crew on a site where the hazards are considered low would not require the same length orientation as a nonexperienced crew at a site they had never seen or heard of. Initial orientation on a large, complicated or extremely hazardous site may take several hours or up to a full day or more. The HAZWOPER standard calls for 3 days of on-the-job training by a qualified person before allowing a worker to be considered "qualified" [1].

7.2 FOLLOW-UP

After workers have completed orientation, the next step is to make sure that the rules or guidelines set up during orientation become a reality in the field. To accomplish the field reality, it is going to take follow-up in the field. Someone or some group of persons will have to leave the office or office trailer and perform field inspections. The field inspections should begin immediately. If possible, we recommend giving the newly hired worker a couple of hours to prepare and get familiarized with the surroundings.

If during the orientation all workers are notified that certain areas should be avoided because of the potential for injury, the penalty for failure to comply should be communicated to everyone. If someone has been noncompliant, communicating how the noncompliance was addressed can aid in avoiding the same situation in the future. Although we believe in being positive, enforcement has its place. Enforcement becomes the key when certain areas have limited access due to potential hazard.

7.3 INSPECTION PROGRAM

Although inspecting the worksite is important for enforcement of important requirements, it is also a useful tool to help determine if the site orientation, the safety plan, or the safety program is effective. If newly oriented workers are out of compliance in certain areas, this may indicate that the orientation needs to be reviewed and improved.

The audit/inspection form that you should use can be developed from the safety plan. A qualified person should examine the safety plan and come up with a checklist that should serve as an audit/inspection form. Allowances should be made to include items not specifically noted in the safety plan but that may be observed during field walk-throughs. Certain highly pertinent sections of what OSHA uses when performing a compliance inspection of hazardous waste sites is included in Appendix D. This inspection/audit form covers many of the basics and can be used a general guide.

If field inspections note shortcomings or noncompliance, a system should be set in place to address these issues. We should keep in mind some basic principles. A meeting should be held to discuss and agree that findings from the inspection are valid. After this step, you might also agree on a priority list. Depending on the size of the site, this list may have three or more tiers. The most important items would be set highest on the top tier of the priority list. Those with low priority would be in a group in a lower hierarchy. Each item is assigned an owner. The owner should be invited to participate in agreeing to an acceptable completion date. Records will be kept to track progress and completion of items. In addition, site workers will be communicated with as to progress.

If your safety plan is comprehensive, it should specify defined roles and responsibilities. The safety plan will state what procedures should be followed when workers come upon a safety-related situation that they cannot "fix" themselves.

If workers are observed working in an unsafe manner, this might indicate a lack of training or qualifications. Workers should never attempt to perform work for which they are not fully qualified. This point cannot be stressed enough. Many accident investigation reports indicate that the lack of fully trained and qualified workers is a root cause or underlying factor in a serious incident.

Communication is the key. Just as orientation is the first thing we do to communicate the safety plan, communicating results of inspection is another important facet of ensuring a safe site. Communication can be completed in a variety of ways. We believe bulletin boards placed in lunch rooms, hallways, or meeting areas can be very effective. E-mails can be used but are less effective if the people you are trying to reach do not have an e-mail address, do not check their e-mails regularly, or are not confident about using computers in general. However, using e-mails to communicate to people other than those already specified can be a great way of informing parts of the team.

As previously mentioned, it is common practice for the same person to wear many hats for smaller, less-complicated sites. The field inspection might indicate that the person performing the double role of SSHO and SS cannot adequately perform the required job functions for both jobs. If this is the case, arrangements should be made to bring in additional personnel and management support.

7.4 JOB HAZARD ANALYSIS

Our reaction should include a new look at the unexpected work activity. One effective way to do this is through a job hazard analysis. Job hazard analysis was discussed in detail in a previous chapter, but for now, keep in mind that when the unexpected occurs you should react quickly and get the whole team involved. Make sure to include the field supervisor and worker.

The hazard analysis is often referred to as hazard characterization. No matter what terminology is used, the idea is to determine what the hazards affecting the worker are and then ensure that adequate controls are in place to protect the worker. Figure 7-1 depicts the process which begins with hazard characterization and goes on indefinitely. The reason that the process never ends is due to periodic reevaluation or reassessment of the hazard. This reassessment ensures that the controls being used still provide effective protection. And as we have mentioned in previous chapters, monitoring the safety program for continued effectiveness is part of the HAZWOPER standard, and believed by the authors to be good business practice.

7.5 TEAM MAKE-UP

In the last chapter we discussed the importance of well-defined roles. This holds true for the inspection team. For many larger sites a union contract may exist that may specify who participates in the inspection/audit process. At smaller sites, this may be open-ended. We believe that the personnel make-up of an inspection team should depend on the size, complexity, number of employees, and on-site hazards at the site in question. Again, preplanning coupled with a common-sense approach should be the driver.

The team should consist of members such as the PM, SS, safety department, training, maintenance, research, and however many others make sense for your site. If your site includes a building with a lot of activity, a representative who works in that building might be asked to participate in an inspection/audit. The team that audits might contain a variety of temporary participants.

For example, if subcontractors are being utilized for the performance of a certain phase of work and will be on site only temporarily, a representative from the subcontractors might be asked to temporarily participate on the team. Even if each subcontractor does not have representation on the team, they should all be given results from the last audit and a contact person to notify if they observe anything that the team might find noteworthy.

An effective inspection program is an integral part of promoting a safe worksite. If you can carefully choose your team members you should be able to produce a document that will guide site activities toward safe work practices. This process will start with the development of an effective HASP document. Once the HASP is developed, it is communicated to site workers and management via orientation and training sessions. The inspection program makes sure that the principles outlined in the HASP are enacted. Should the inspection process determine shortcomings, adjustments must be made to address those shortcomings. The adjustments may be indicated within the HASP, training, orientation, or other areas to provide for a safe worksite. Keep in mind that the

Implementing the Safety Plan 93

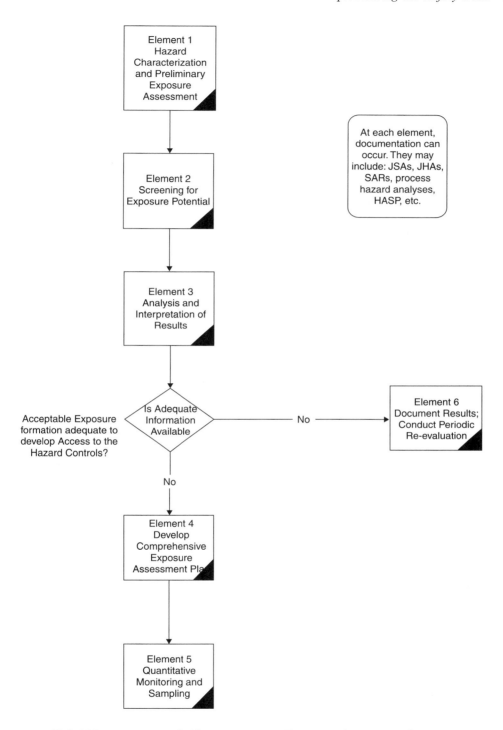

FIGURE 7-1. *Hazard Characterization Exposure Assessment Strategy*

HAZWOPER standard requires that the program be monitored to determine its continued effectiveness [2]. The inspection process is an excellent tool in determining program effectiveness.

7.6 ASSESSING PPE

Any time PPE is used, its proper use should be assessed. An effective inspection process can help identify problem areas with PPE. Observations can be made to visually inspect PPE as the walk-through is being conducted. First, you should determine if the PPE observed what has been specified in the HASP. Second, determine if you have observed instances when the PPE was overprotective, underprotective, or inadequate in any way.

Finding the right types and amounts of PPE provides a constant challenge to the safety professional. Most safety people have a tendency to overprotect workers. After all, we would rather be safe than sorry. However, many times this point is taken to an extreme. The modern challenge is how to adequately protect workers but not overprotect them.

Overprotection regarding the use of respiratory protection has been well documented. We will discuss the dangers of respiratory protection in detail later in this book. Less well known than over-respiratory protection are the dangers of overprotection with other types of PPE. For example, the requirements of coveralls, saranex, or other whole-body protection during hot weather can be a tremendous stressor. Since heat stress comes into play, safety professionals will counter this by adding a cooling vest. If you have ever worked with cooling vests you understand that they add considerable weight and only cool the midsection of the body. What can end up happening is that the site heat stress program will call for a work/rest regimen to be put into effect. This heat stress program may allow only fifteen minutes of work activity for every hour of duty (45 minutes of rest). When this occurs, it typically causes extra manpower to be utilized, and drives up costs considerably.

Overprotection can also be detrimental to the hearing protection program. As an example, let's consider the following situation. For ease of administration or zealousness, a company institutes a universal safety rule stating that "hearing protection will be worn at all times." Although management may have the best of intentions in trying to protect worker hearing, wearing hearing protection when not required can have detrimental effects. Some of these effects include:

- Inability to verbally communicate
- Inability to hear sounds at different noise levels or frequencies
- Discomfort

If health and safety professionals were able to place every worker in a protective bubble where the atmosphere was controlled and no injury or illness could befall the worker, this would make the world a safer place. This bubble is not yet a reality; however, we believe that using an exorbitant amount of PPE satisfies the safety professional's desire to place workers in a protective bubble. Until this bubble becomes reality, we should consider some practical alternatives. These include:

- Scheduling work during the time of the day or the season of the year that minimizes heat/cold stress potential
- Using remote equipment
- Using robots

The safety professional should always be inviting ideas to make sure that site workers are working smarter and not harder. Refer to Chapter 9 for more information on PPE.

REFERENCES

1. 29 CFR 1910.120 (q).
2. 29 CFR 1910.120 (b) (4) (IV).

Chapter 8

Training Requirements

Effective training is one of the most important keys to worker safety and health. Training represents a significant portion of the cost of implementation of the hazardous waste standard, and is important anytime when working with hazardous materials. Training requirements under HAZWOPER, 29 CFR 1910.120 has a major impact at all hazardous waste sites such as DOE, the Army Corps of Engineers, and other related sites. We need to keep in mind that training workers is, plain and simple, a good management practice. This is true whether or not a worksite or work activity is specifically covered by the standard. Having workers who are qualified to perform work activities is just a basic necessity.

This chapter is intended to provide the PM with guidance that can be used for implementing training. We continually refer to hazardous waste training requirements because we believe that the underlying principles are applicable to all situations that deal with hazardous substances. Obviously, if one is dealing with a regulated site, the hazardous waste standard and other requirements should be integrated into the training program as applicable.

8.1 SYSTEMATIC APPROACH TO TRAINING

DOE recommends the use of a "systematic approach to training," in which the content of training is commensurate with the potential hazards, exposures, worker roles and responsibilities, and requirements of the project (see Figure 8-1) [1]. The description of this systematic approach sounds like a great idea. However, in some cases the execution of the systematic approach is difficult to attain. In general, training classes aim content and level to reach at least 80 percent of attendees.

Training for other activities such as deactivation and D&D may not fall under the hazardous waste definition. As previously mentioned, the authors believe that, in many cases, applying hazardous waste principles based on a hazard-based approach will help to provide a safe worksite and add value to these activities. These activities may involve hazard abatement processes, such as chemical lab packing, asbestos, lead, mercury, or

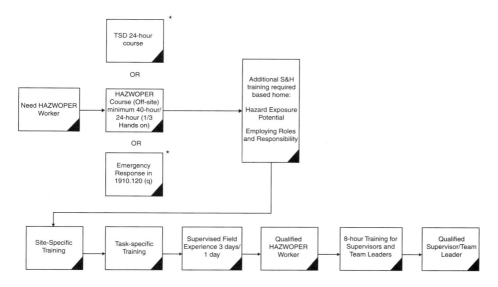

FIGURE 8-1. *Site Worker Training Flow Diagram*

beryllium abatement, and radiological decontamination. Safety hazards can involve the whole gambit of issues. As your experience level increases, the types of hazards encountered will likely increase. These hazards may include, but are not limited to, issues related to construction, confined-space entry, lockout/tagout, hoisting and rigging, and use of elevated platforms or forklifts. Training for these requirements should be based on the hazards of the activity. After the appropriate training requirements are defined, they should be outlined in the site-specific safety plan.

Let's consider lead abatement or asbestos work. These activities provide a good example of how hazards are minimized by controlling access. When working with either substance, an enclosure can be constructed that keeps out unauthorized people and contains the hazardous substance. The only persons who should be potentially exposed are those who are trained, qualified, and medically fit personnel who deal appropriately with the hazard. Workers in the enclosure are protected by PPE, respiratory protection, engineered ventilation systems such as negative air machines, high-efficiency particulate air (HEPA) vacuums, and administrative controls.

8.2 GENERAL TRAINING REQUIREMENTS AND GUIDELINES

Under the hazardous waste standard, paragraphs (e) and (p) specify training requirements for employees who may be exposed to health and safety hazards at cleanup sites and Resource Conservation and Recovery Act of 1976 (RCRA) TSD facilities, respectively. Paragraph (q) specifies training requirements for employees who participate in emergency responses at locations other than cleanup sites and RCRA TSD facilities. Even if the site under consideration is not covered by the above requirements, the appropriately trained workers will be a key in the safe and efficient performance of work tasks.

29 CFR 1910.121, "OSHA Accreditation of Training Programs for Hazardous Waste Operations" (proposed) and the nonmandatory Appendix E to the HAZWOPER standard, "Suggested Training Curriculum Guidelines," are recommended for site-specific implementation. These nonmandatory guidelines provide a common-sense approach to help management choose the appropriate programs. When considering an outside contractor, you should always include logistics. This part of the selection process is important. Very simply put: "Can the outside training contractor provide my workers with instruction that is convenient for workers to attend and that will be completed before the work tasks begin?"

After the basic need for logistics has been met, the next and most important step should be considered.: "Will workers receive quality training and be provided the information they need in a format that they will retain and use?" If these two basic needs are not met, more careful consideration and research need to be implemented. To assist in making this determination, the nonmandatory requirements already mentioned should prove helpful. The types of subjects that are discussed in these nonmandatory appendices include:

- Experience of the instructors
- Course curriculum
- Agenda
- Testing
- Location, size, and condition of the training classroom
- Audiovisual materials that will be used
- The number of similar classes
- A client list

If would be advisable to review the program by touring the training facility and meeting the instructors. It would also be beneficial if you could attend a similar class that is being taught. This would help you to judge the quality of instruction.

Another reference that you may find helpful includes 29 CFR 1926.65 Appendix E (nonmandatory).

8.3 SUPERVISED FIELD EXPERIENCE

General site workers who attend the 40-hour course must have a minimum of 3 days of supervised field experience under the direct supervision of a trained, experienced person prior to being qualified to work unaccompanied. Workers who receive the 24-hour course are required to have 1 day of supervised field experience. If an employee changes tasks and the work is significantly different, all or part of the supervised field experience may need to be repeated, even on the same hazardous waste worksite.

The primary intent of supervised field experience is for employees to be observed by the experienced person during the course of their workday to ensure that they are working safely. These 1- or 3-day observation periods allow the experienced person to observe the worker applying proper techniques and to emphasize site-specific hazards and special working conditions. The observation period includes some on-the-job training as a reinforcement of previous training received (see Figure 8-2) [1].

Any designated, trained (8-hour HAZWOPER supervisor course as a minimum), and experienced individual responsible for the safety of an employee (such as team leaders or crew leaders) may perform the function of an experienced person to provide the "supervised field experience" required by HAZWOPER [1]. Although having the appropriate certificates of completion would satisfy regulatory requirements, you should also consider time of service and experience. A fresh college graduate with training certificates and minimal field experience may be less desirable to perform supervised field experience than the safety professional with years of substantial field experience.

8.4 TRAINING CERTIFICATION

Initial hazardous waste training certification depends on two criteria:

- The successful completion of a 40- or 24-hour training course
- Completion of the specified level of supervised field experience

The employer is responsible for making sure that both requirements are met before final certification is granted.

Worksite-specific scenarios and hands-on use of equipment should be included as much as possible in training (recommended minimum of one-third of course hours). Specific examples should be used in all courses. Likewise, any discussion of hazards should include site-specific hazards. Special consideration is warranted for providing practical, hands-on training for emergency responders; emergency response training typically involves practice drills and demonstrations. The state fire marshal or authority having jurisdiction should be consulted to make certain that HAZWOPER, DOE, or Corps of Engineers training requirements for emergency response are met including any state- or community-specific requirements.

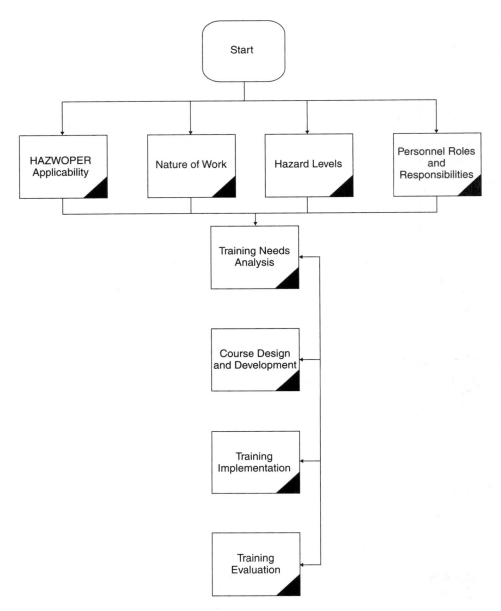

FIGURE 8-2. *Systematic Approach to Training*

8.5 SPECIFIC TRAINING GUIDELINES

Prior to beginning any training activity, exact training needs should be identified. Training needs may vary based on hazards, potential exposures, work requirements, roles and responsibilities, job descriptions, and compliance requirements. Job hazard analyses and employee surveys are some of the tools used in determining specific training needs (see Figure 8-2).

8.6 INSTRUCTOR/TRAINER QUALIFICATION

Instructors providing training need to be qualified in their areas of instruction. This qualification is based on documented experience, successful completion of a "train-the-trainer" program, and an evaluation of instructional competence. Instructors should maintain professional competency by participating in continuing education or professional development programs or by successfully completing periodic instructor refresher courses and competency reviews.

8.7 PROGRAM AND COURSE EVALUATIONS

Training programs and courses should be monitored and revisions made by training or environment, safety, and health professionals as a result of comments provided by the students, other instructors, and supervisors. Training should also reflect changes in policies and procedures, site characterizations, job requirements, lessons learned, and regulatory requirements. Adjustments should be made as a result of analyzing work experiences at similar sites and based on accepted good practices.

8.8 EMERGENCY RESPONSE TRAINING

Under the hazardous waste standard, on-site emergency responders, on-site collateral-duty emergency responders, and off-site emergency responders are trained to one or more of five levels of competency, depending on the type of response they perform as specified in 29 CFR 1910.120 (q).

- First responder awareness level
- First responder operations level
- Hazardous materials technician
- Hazardous materials specialist
- On-scene incident commander

Beyond these five key levels, there are two specialized categories of emergency responders: skilled support personnel and specialist employees.

There are five general categories of emergency response personnel who respond to specific types of emergencies:

- Hazardous waste cleanup site workers who respond to emergencies in addition to their normal duties at the site they are assigned to, as specified in 29 CFR 1910.120 (l) and (e)
- TSD facility workers who respond to emergencies in addition to their normal duties at the facility they are assigned to, as specified in 29 CFR 1910.120 (p)
- On-site collateral-duty emergency responders who respond to limited emergencies on an as-needed basis within a defined work area, as specified in 29 CFR 1910.120 (q)
- On-site emergency responders who respond to emergencies regardless of type or location on a full-time basis, as specified in 29 CFR 1910.120 (q)
- Off-site emergency responders who respond to emergencies on a full-time basis regardless of type or location of the emergency, as specified in 29 CFR 1910.120 (q)

The latter three categories include all responders not covered by 29 CFR 1910.120 (l) and (p). The last category includes local firefighters and HAZMAT teams (see Table 8-1).

HAZWOPER establishes five categories of training requirements based on the duties performed by emergency responders. OSHA determined that job responsibilities define training requirements but that training does not define job responsibilities. When working with hazardous materials, and especially during emergencies, not only the tasks performed but who performs them is of the utmost importance. If responders have not been trained in a specific task and informed that they will perform the task during response, they are not permitted to perform the task regardless of their training level (see Table 8-2).

8.9 LESSONS LEARNED

Extremes appear to be commonplace in the hazardous waste industry. Although not common on large projects or DOE sites, workers without 40-hour training have been performing work for which 40-hour training is required. Every individual should be encouraged to keep copies of all successfully completed training. Wallet cards are encouraged. Employees can keep these cards with them at all times. They allow employees to monitor their own training compliance while offering some proof to auditors or regulators that the worker has been appropriately trained.

When interviewing potential workers for hazardous waste work, the interviewer should determine if the interviewee is up to date in training. Should the perspective worker get the job, how much time, effort and money will it take in training to get that worker up to speed? It is

TABLE 8-1 Emergency Responder Categories and Training Requirements on a DOE Site.

Category	Training Criteria	Definition	Refresher Training
Hazardous Waste Cleanup Site Workers Who Respond to Emergencies	29 CFR 1910.120 (e) 29 CFR 1910.120 (l)	Individuals working at a hazardous waste cleanup site who respond to emergencies in addition to normally assigned duties	Rehearse emergency plan in annual 8-hour refresher training
Treatment, Storage, and Disposal (TSD) Facility Workers Who Respond to Emergency	29 CFR 1910.120 (p)	Individual working in TSD facility who responds to emergencies in addition to normally assigned duties	Rehearse emergency plan in annual 8-hour refresher training
On-Site Collateral-Duty Emergency Responder	29 CFR 1910.120 (q)	Individual in a work area who is trained to respond to limited emergencies on an as-needed basis: not a full-time responder	Practice and drills as necessary
On-Site Emergency Responder	29 CFR 1910.120 (q)	Full-time emergency responder on DOE site who responds to emergencies at the site	Practice and drills as necessary
On-Site Emergency Responder	29 CFR 1910.120 (q) or state mandate*	Personnel from outside DOE site who respond to an emergency on the DOE site. Includes local fire fighters, HAZMAT teams, etc.	Practice and drills as necessary.

*State and local employees are not covered under the OSHA Act, HAZWOPER, or other OSHA regulations, but are often covered by state safety and health regulations.

Modified HAZWOPER Categories and Training Requirements for Emergency Responders Adopted from U.S. Department of Energy *Handbook for Occupational Safety and Health*, June 1996, pp. 4–12.

TABLE 8-2 Summary of Training Requirements for Emergency Response Personnel.

Job Title	Definition	Training Requirements
Skilled Support Personnel	• Expert in the operations of equipment, not necessarily employees of the employer, and may perform temporary emergency response • Examples: crane or earth-moving equipment operations, or medical personnel whose typically duties do not include treating contaminated patients.	• Must receive initial briefing at the site prior to participation in emergency response as required by 29 CFR 1910.120 (q)(4) • Demonstrated competencies
Specialist Employees	• Employees outside immediate release area who assist on-scene incident commander • All activities coordinated through individual in charge of the incident command system • Examples: industrial hygienists or health physicists providing guidance on PPE selection	• Must meet requirements of 29 CFR 1910.120 (q)(5) • Demonstrated competencies listed in NFPA Standard 472, 1992 for specialist categories C, B, and A
First Responder Awareness Level	• Individuals likely to witness or discover a release and who are trained to initiate emergency response sequence • Example: security personnel	• Must meet requirements of 29 CFR 1910.120 (q)(6)(I) • Demonstrated competencies listed in 29 CFR 1910.120, Appendix E
First Responder Operations Level	• Individuals who respond to releases in a defensive fashion and confine it from a distance • Example: firefighters, since they will respond to releases; and process operators, since they may take defensive actions from a safe distance	• A minimum of 8 hours of training or sufficient experience to demonstrate competency in areas listed in 29 CFR (q)(6)(ii)[1] • Demonstrated competencies listed in 29 CFR 1910.120, Appendix E

Training Requirements

HAZMAT Technician	• Responds to releases for purpose of stopping release • Process operators may take limited action in danger areas if they: (1) have informed the incident command structure of the emergency, (2) have adequate PPE, (3) have adequate training in procedures they are to perform, and (4) employ the buddy system	• A minimum of 24 hours of training or sufficient experience to demonstrate competency in areas listed in 29 CFR (q)(6)(iii)[2] • Demonstrated competencies listed in 29 CFR 1910.120, Appendix E
HAZMAT Specialist	• Duties parallel HAZMAT technician's • Requires knowledge of substances to be contained	• Includes demonstrations and hands-on performance and proficiency • At least 24 hours of training equal to the HAZMAT technician level and additional competency in areas listed in 29 CFR (q)(6)(iv)[2]
On-Scene Incident Commander	• Oversees HAZMAT team and is knowledgeable in command and management • Does not necessarily have extensive knowledge of certain technical aspects	• A minimum of 24 hours of training equal to the first responder operations level and additional competency in areas listed in 29 CFR (q)(6)(v)[2] • Demonstrated competencies listed in 29 CFR 1910.120, Appendix E

1 California State Fire Marshal's Office and other states require 16 hours of training.
2 The California Office of Emergency Services requires 160 and 240 hours of training for HAZMAT Technician and Specialists, respectively, for state certifications. However, state certifications for HAZMAT Technicians and Specialists is not required.
Note: It is important to determine state and other requirements in your jurisdiction.
Adopted from U.S. Department of Energy *Handbook for Occupational Safety and Health*, June 1996, pp. 4–6.

important not just to consider the cost of coursework, but the amount (in salary) of time needed to complete the coursework.

All types of businesses are interested in cutting costs. You should not be surprised when, during an interview, an experienced worker informs you that his or her present or past employers did not make the appropriate investment in them as far as training is concerned.

Is there such a thing as "over"-training? Most persons would agree that there is no such thing as too much training. However, balance needs to be mentioned. Usually, some type of a cost justification is performed prior to sponsoring an employee's training. This is particularly true if the course involves distant places and expensive tuition.

Training is not always given just because it is required. Sometimes employees attend training as a perk. There may be situations when forcing an employee to attend training at a certain time or location might be considered a punishment. Worker morale can suffer when workers are forced to attend training sessions just because "everyone must attend." Many times, the worker cannot understand why he or she has to attend the training. The workers may feel that the training in question does not apply to their job class or the training may involve subjects that are foreign to them. In either case, the worker feels forced to attend training. Forcing a worker to do something, even if the employer has good intentions, is typically less advantageous to obtaining worker buy-in. Reasonable steps should be taken to promote worker buy-in and minimize the use of force.

Mandatory training, even when there is worker cooperation and buy-in, should be carefully considered. Mandatory training has certain advantages over spot training. Let's take a situation where a certain amount of specialized training is required to perform a lucrative client service that your firm performs. The management of the firm has found it advantageous to have everyone cross-trained so that any available person can perform this lucrative service. The training is mandatory, and every new hire is scheduled to complete this training. Companies find that they can satisfy client needs by making sure that any worker available can perform the specified lucrative service. Scheduling is less difficult; record keeping and finding qualified workers are also less difficult. Finding the qualified worker to fill in when another is sick or on vacation is easy. Under these circumstances, workers should easily understand why they are required to complete this training.

REFERENCE

1. *Handbook for Occupational Health and Safety During Hazardous Waste Activities.* Office of Environmental Safety and Health Office of Environmental Management, 1996, pp. 4–5, 4–9, 4–10.

Chapter 9

Personal Protective Equipment

PPE is an important part of working with hazardous materials, and is used and accepted in many situations at home and at work. As you may remember from previous chapters, you should use engineering and administrative controls before you use PPE. Although PPE has come of age, PPE is still to be considered the last line of defense in the prevention of accidents.

The emergence of designer-type safety glasses is an example of how PPE has become part of nearly everyone's life. Home improvement and hardware stores typically pick a spokesperson who just happens to be wearing designer safety glasses. Another example of the acceptance of PPE is back supports. Ten years ago if you went to an airport where a limousine driver was loading or unloading luggage, you would be hard pressed to find anyone wearing a back support. In my most recent trip to the airport, more drivers than not were wearing back supports. Is this a good source of back injury prevention? Some people think back supports are effective; many remain non-believers.

There are many schools of thought on the use of PPE. We will outline and define some of the most important PPEs that should be used when dealing with hazardous materials.

9.1 GENERAL USAGE OF PPE

The use of proper PPE is an integral part of many jobs when dealing with hazardous waste. OSHA standard 1910.132 of 1998 requires an assessment be conducted to determine the appropriate PPE for eyes, face, head, and extremities whenever hazards encountered are capable of causing injury or impairment in the function of any part of the body through absorption, inhalation, or physical contact. According to the PPE standard, it is the employer's responsibility to determine if hazards are present (or likely to be present). If the employer determines that hazards are present, the employer should choose the types of PPE that will protect affected employees from the hazards identified in the hazard assessment [1].

In addition to assessing the hazard and choosing the PPE, the employer should communicate selection decisions to all affected workers and train them so that they thoroughly understand the requirements of the selected PPE. Once the training has been completed, the employer should verify through a written certification what type of training has been completed [1].

9.2 SELECTING PPE FOR HAZARDOUS WASTE ACTIVITIES

In the hazardous waste environment a level of protection should be specified for each job task, as appropriate. For nonhazardous waste jobs the levels may not be recognized or accepted. In this section we will first discuss PPE for hazardous waste activities and then discuss PPE in general.

The PM and SS should be aware that no single combination of PPE can guard against all hazards because every worksite is different and the degree of hazards (known or unknown) may vary day by day. The PPE ensemble probably will be required to change as work continues.

Notice that each level of protection specifies a complete clothing ensemble. However, in practice, the level of protection selected for a particular task is driven by the respiratory protection requirements. Once respiratory protection is chosen, clothing is matched to the dermal and safety hazards. OSHA requires that the level of PPE be selected based on three distinct tasks:

- Conducting a hazard characterization and exposure assessment to identify:
    Actual or potential hazards
    Possible exposure routes
- Organizing and analyzing the data and selecting PPE based on the type of hazards, the level of risk, and the seriousness of potential harm from each identified hazard
- Making sure that the level of PPE selected fits properly and protects the user against the hazards
- Periodically reassessing the hazards and PPE selection [2]

Manufacturers' literature is often the best source of information for selecting PPE. To help with appropriate PPE selection there are some other useful references that are readily available.

- *Guidelines for the Selection of Chemical-Protective Clothing* by J.J. Johnson and A.D. Schwope et al., published by the American Conference of Governmental Industrial Hygienists
- *Standard Operating Safety Guides*, published by the U.S. EPA Office of Emergency and Remedial Response

- *Occupational Safety and Health Guidance Manual for Hazardous Waste Site Activities*, published by National Institute for Occupational Safety and Health (NIOSH), OSHA, the U.S. Coast Guard, and U.S. EPA.

The cited references provide additional and more detailed information on issues such as advantages and disadvantages of PPE, compatibility of various types of PPE with chemical hazards, respiratory protection factors, training and proper fitting, and consideration of work duration.

For hazardous waste activities, the levels of protection can be classified as four groups: Levels A, B, C, and D. Level A is the most protective and level D the least protective. The following examples outline the typical types of protection.

9.2.1 Level A

This type of protection offers the highest protection in regard to respiratory, skin, and eye protection. This level may consist of the following elements:

- Pressure-demand full-facepiece self-contained breathing apparatus (SCBA) or supplied-air respirator (SAR)
- Fully-encapsulating chemical-resistant suit
- Inner chemical-resistant gloves
- Chemical-resistant safety boots
- Disposable gloves and boot covers (these are worn over the encapsulating suit)
- Coveralls
- Hard hat

Recommended:

- Long cotton underwear (dependent on site conditions)
- Two-way, voice-activated radios
- Cooling units (dependent on site conditions) [3]

Level A suits limit personal mobility. There is a drag on every joint in your system. Your range of motion is limited. This drag on your body in general contributes to the inability to perform work tasks. Level A suits are constructed to fit a wide user group. The amount of material used coupled with the shape of the suits will usually fit people who may be robust or out of good physical condition. The extra material around the midsection can be cumbersome and impeding for someone who is not overweight. Moreover, for the worker who is overweight, there is typically never enough material to get the proper fit.

Figure 9-1 is an example of what a modern level A ensemble can look like. Remember, level A suits are fully encapsulated. The ensemble would come with gloves and boots included. An overboot and overgloves would be added to provide extra protection and add extra working life to the ensemble. An SCBA is worn with the air pack unit fully within the ensemble, and the worker wears a full facepiece. Notice the field of vision that the worker enjoys with the ensemble shown. Were the ensemble in Figure 9-1 not fully encapsulated at the sleeves, feet, or seams, it would not be considered level A.

Fully encapsulating suits do not contain speaking diaphragms. You can attempt to yell through the suit at your partner, but this usually proves ineffective. Hand signals and an agreement as to their meaning can come in very handy. Voice-activated radios seem to be the best alternative at this time. However, all workers should be well aware of hand signals if the radios fail. The inside of a level A suit is quite loud. The noise comes from the materials as the person inside the suit moves and from the inhalation and exhalation of air through the SCBA. This noisy environment makes it difficult to hear potential danger sounds such as hissing drums or heavy equipment. Obviously, this lack of communication and the inability to hear is a potential safety hazard.

When I first put on the level A suit and turned my head, I was quite surprised when the window did not turn with me. The result was a close look at the inside of my suit. I quickly learned to turn my whole body in the direction that I wished to look. Attempting to look up and down provided the same challenges. When I attempted to look up or down, I realized quickly that, if I wanted to see anything, I should keep my neck "stiff." If my neck was not stiff, I could not see the work area. We believe that the lack of vision is the single greatest challenge that workers face while working in level A protection.

Working in level A protection can cause a variety of stresses. The equipment is heavy. The pressure to complete work tasks during a time frame is intensified because work time is limited by air supply. Heat stress can be a problem, even in the winter. In the summer, the use of cooling vests can keep you cool but also adds to the weight that you are carrying. Typically, all level A workers have a sharp knife blade so that they can cut themselves out of the suit if the air supply fails. Realizing that you may have to cut yourself out of this suit in case of air supply failure adds more potential stress to the situation.

Adding to the general stress is a fear factor. Some level of fear stems from a knowledge that the materials you are dealing with are dangerous and the only way to be adequately protected is to wear a totally encapsulated suit. The use of any other lesser protection could cause the employee harm. The fear factor is compounded by the fear of running out of air and not being able to cut yourself out of the suit in time before

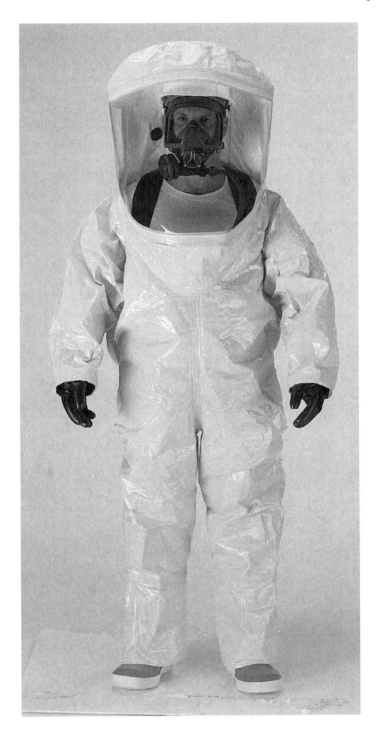

FIGURE 9-1. *This worker has donned a modern, full-body protective garment with a virtually unobstructed view. Photo courtesy of DuPont Tyvek®/ Tychem® protective apparel*

you pass out. If you were able to cut yourself out of the suit, what type of hazard would you be exposed to?

The general stress coupled with all of these disadvantages soon made the level A protection rare. The rare use of level A equipment has trickled down to training courses, where it is increasingly difficult to find a course curriculum that requires level A training.

In Figure 9-2 we see a worker demonstrating a level A ensemble handling a drum. Notice that his SCBA airlines and other apparatus are covered and protected by a layer of clothing.

In Figure 9-3 we see two workers (demonstrating the "buddy system") working near a tank farm and caged ladder. This type of ensemble can be a distinct advantage because the suit covers all of the workings of the SCBA and makes it less likely that hoses or apparatus could get caught on anything while performing work activity.

FIGURE 9-2. *This worker is shown handling a placarded drum. Photo courtesy of DuPont Tyvek®/ Tychem® protective apparel*

Personal Protective Equipment 113

FIGURE 9-3. *These two workers appear to be working on a tank farm. One of the workers appears to be descending a caged ladder. Note how his airlines and apparatus are not likely to get tangled in the cage. Photo courtesy of DuPont Tyvek®/ Tychem® protective apparel*

9.2.2 Level B

This level has the same respiratory and eye protection as Level A but requires less skin protection. This level may consist of the following equipment:

- Pressure-demand full-facepiece SCBA or SAR
- Chemical-resistant clothing
- Inner and outer chemical-resistant gloves
- Chemical-resistant safety boots
- Disposable boot covers
- Coveralls
- Hard hat

Recommended:

- Long cotton underwear/cooling vests (dependent on site conditions)
- Two-way voice-activated radios [3]

Notice the gloves that the worker in Figure 9-4 is wearing and the field of vision. This is a typical worker performing in level B protection. You might also find that workers will duct tape the chemical protective gloves to their clothing as a way to further protect against leakage contacting their arms or body. You might also find underneath the gauntlet gloves that the worker is wearing latex gloves that are duct taped to the fabric for extra protection. This suit is not fully encapsulated, and the SCBA is not protected by the protective suit.

In Figure 9-5 the worker is ascending a caged ladder. Notice that the worker's air pack, airline or apparatus could become entangled with the ladder protection. The worker in Figure 9-6 who is using absorbent to soak up a mock spill has no encumbrances pictured near the work area. So, besides the proper level of protection, the type of work being performed and the work area can be important.

Workers pictured in Figure 9-7 are wearing suits that cover and protect the components of their SCBAs. This can be important when carrying objects that can catch or obstruct or when working near stairways or ladders. Depending on whether or not these suits are fully encapsulated (checking how the sleeves and feet are made), these ensembles could be either level A or level B.

Workers shown in Figure 9-8 are demonstrating a typical level B ensemble. They are also practicing the buddy system. Take heed that their airlines might become entangled when using the ladder to cross the diked area.

9.2.3 Level C

This level includes hazard-based skin and eye protection but less respiratory protection than Level B. This level may consist of the following equipment:

- Full-facepiece air-purifying respirator (APR)
- Chemical-resistant clothing
- Inner and outer chemical-resistant gloves
- Chemical-resistant safety boots
- Disposable boot covers
- Coveralls
- Hard hat

FIGURE 9-4. *This worker is using typical level B protection while handling a drum. Photo courtesy of DuPont Tyvek®/ Tychem® protective apparel*

116 *Hazardous Waste Compliance*

FIGURE 9-5. *This worker appears to be ascending a caged ladder. Note the likelihood of airlines or apparatus becoming entangled in the cage when the worker is on the descent. Photo courtesy of DuPont Tyvek®/ Tychem® protective apparel*

Recommended:

- Long cotton underwear/cooling vests dependent on site conditions
- Two-way voice-activated radios [3]

We should keep in mind that the main difference between level D and C equipment is the amount of respiratory protection. Level C protection requires use of a respiratory device (APR). Requiring workers to work in level C for more than a small percentage of the time can prove to be a challenging situation for both workers and managers. Respirators, especially full-face respirators, can provide excellent protection for workers, but are also found to be a source for worker complaints.

Respiratory protection should always be carefully considered by a qualified person who is aware of the specific task and site conditions. Similar stresses as those pointed out for Levels A and B can be found

Personal Protective Equipment 117

FIGURE 9-6. *This worker is shown handling spill cleanup in level B protection. Photo courtesy of DuPont Tyvek®/ Tychem® protective apparel*

for level C, although usually not to the same extent. Asbestos abatement workers might typically wear respiratory protection while performing abatement activities.

Figures 9-9 and 9-10 show workers wearing typical modified level C protection while performing different work activities. Notice the extra mobility that goes with decreasing level of protection.

9.2.4 Level D

There is no respiratory protection required for this level. There is minimal skin protection due to contact. This level may consist of the following equipment:

- Coveralls
- Abrasion-resistant gloves

118 *Hazardous Waste Compliance*

FIGURE 9-7. *These protected workers are demonstrating the buddy system. They appear to be trying to communicate with each other. Hand signals or communication should be agreed on prior to the inception of work activities. Protective clothing can inhibit communication. Photo courtesy of DuPont Tyvek®/Tychem® protective apparel*

- Safety boots
- Disposable boot covers
- Hard hat
- Face shield (for flying-debris hazards)
- Escape mask [3]

In Figure 9-11 two workers in level D protection are pictured overseeing work activity on a platform. One appears to be a supervisor, the other a worker. Both are wearing level D, but the whole body protection and hand and head protection are different.

FIGURE 9-8. *Two workers in level B are practicing the buddy system. Photo courtesy of DuPont Tyvek®/ Tychem® protective apparel*

9.2.5 Modified Level D

This type of protection would require some additional protection that a basic Level D does not provide. For example, if you are working at a truck stop where there is a high level of traffic, you may need a traffic vest. In another example you might be working over water. In that situation you may need a floatation vest certified by the Coast Guard. Other items considered modified level D include: safety harnesses, lifelines, vibration cushioning gloves, and electrical lineman's gloves. This list could be expanded based on the task and hazard control chosen.

The levels of PPE discussed provide controls of the hazardous substance based on the degree of worker exposure. As we have discussed before, PPE is only acceptable as a hazard control measure in the following situations:

FIGURE 9-9. *A level C worker is moving a drum with a drum dolly. Photo courtesy of DuPont Tyvek®/ Tychem® protective apparel*

- Engineering or administrative controls are not feasible or do not eliminate the hazard
- Engineering controls are being developed
- During emergencies [3]

9.3 UPGRADING OR DOWNGRADING LEVELS OF PROTECTION

The PM, SS, and SSHO are responsible for making a decision for upgrading or downgrading the level of PPE based on provisions specified in the HASP. Clear criteria should be established based on Table 9-1.

There are some additional requirements imposed when respiratory protection is specified. The following are some considerations in

Personal Protective Equipment 121

FIGURE 9-10. *Working on stairs makes work that much more difficult for this worker in level C. Photo courtesy of DuPont Tyvek®/ Tychem® protective apparel*

TABLE 9-1. Upgrading and Downgrading PPE.

Upgrading PPE
- Unstable or unpredictable worksite hazards or emissions
- Known or suspected presence of dermal hazards
- Occurrence or likely occurrence of gas or vapor emission
- Change in work task that increases the potential for contact with hazardous materials

Downgrading PPE
- New information that indicates a situation is less hazardous than orginally thought
- Hazard assessment and monitoring data show low exposure levels
- Change in site conditions that decreases the hazard
- Change in work task that reduces contact with hazardous materials

FIGURE 9-11. *These two workers appear to be looking on from a safe distance. They are both wearing level D protection with minor modifications. Photo courtesy of DuPont Tyvek®/ Tychem® protective apparel*

determining the level of or maintaining a higher level of respiratory protection:

- Working in a respirator can cause unnecessary, potentially dangerous stress to workers.
- Use of respirators limits vision and mobility. This is an important consideration when performing strenuous work activity or when operating heavy equipment.
- Over-reliance on respirators can cause a false sense of security as the protection factor for respirators varies with workplace conditions.
- Implementation of respirator programs is costly.

If worksite hazards have been minimized through engineering and administrative controls, a management decision to use respirators necessitates implementation of requirements mandated by 29 CFR 1910.134 [3].

Special requirements for respiratory protection include the following:

- Preparing a written respiratory protection program, if no written program exists, and appending the new or existing program to the HASP
- Medically evaluating, training, qualifying, and fit-testing workers for specific respirator types, checking 29 CFR 1910, Subpart Z, "Toxic and Hazardous Substances," for any special respiratory protection requirements (e.g., for asbestos, lead, or cadmium) [3]

We discuss respirators as part of PPE later in this chapter. We also provide figures of a variety of different types of respiratory protective devices.

9.4 LESSONS LEARNED REGARDING LEVELS A AND B

Back in the days immediately following Love Canal, workers were typically required to learn Level A protection. My first hands-on training for level A came during my initial 40-hour hazardous waste worker training in 1987. Part of the course requirement included donning the fully encapsulating suit and attempting to play basketball with five other people who were also wearing level A. The effects of this activity are still fresh in my mind.

One of the first things that you will notice when you attempt to perform a task is that you are carrying a lot of extra weight. This extra weight had several effects on me, as I am sure it does to most folks. These effects included:

- Inability to perform work tasks
- Lack of mobility
- Greatly reduced capacity for communication
- Lack of vision
- General stress

Tasks that once were simple, such as bending over or grasping implements, were greatly impeded. Although some tasks could still be performed, the rate at which they were performed was substantially reduced.

9.4.1 More Lessons Learned

Some confusion may exist between non-HAZWOPER and the more traditional hazardous material site. Specifying level A, B, C, or D is confusing for those who have not been trained on hazardous waste. One area of confusion lies in terminology. OSHA has designated a general respiratory classification of type C for air-supplied respirators. Typically, type C respirators are used in level B protection. Level C protection, by definition, would not allow the use of type C respirators. There are a variety of other issues where confusion seems to abound. However, the authors believe that the confusion can be cleared up through adequate training and communication.

9.5 PPE SPECIFICS FOR NONHAZARDOUS WASTE SITES

On April 6, 1994, OSHA published its final revisions to the Personal Protective Equipment (PPE) standard in the *Federal Register*, Vol. 59, No. 66. With the implementation date of July 5, 1994, the regulation, applicable to the general industry, represented major changes in the selection and use of PPE. OSHA believes that through compliance with the PPE standard, safety statistics that track worker safety will improve. These improvements will add up to 712,000 lost workdays and 95,000 recordable cases.

The new standard amended 29 Code of Federal Regulations (CFR) to include general requirements (29 CFR 1910.132), eye and face protection (29 CFR 1910.133), head protection (29 CFR 1910.135), and foot protection (29 CFR 1910.136). A new regulation (29 CFR 1910.138) applied to hand protection. These changes are significant because they mandated employers to conduct a hazard assessment of the workplace to decide if hazards in the operation required the use of PPE.

Employers should provide a written verification that a hazard assessment has been completed. According to the preamble, "benefits will be gained through selecting more appropriate PPE, increased awareness of hazards and improved consistency in use."

9.5.1 General Requirements

Section 29 CFR 1910.132 added new general requirements for the selection and use of PPE to include the following:

- A hazard assessment should be conducted to identify hazards present that would require the use of PPE.
- The appropriate PPE should be selected and properly fitted for each affected employee based on the assessment.
- Defective or damaged PPE should not be used.
- Each employee who is required to use PPE should be trained and retrained as applicable in the proper selection and use of PPE.

- Each employee trained should demonstrate an understanding of the training. The employer should "Certify" in writing that the training was provided and understood by each employee.

The standard does not address the question of who should pay for the required PPE. However, in a compliance memorandum OSHA has clarified its position that in most cases the employer should provide and pay for the employee's PPE. The OSHA memorandum explains, "If the PPE is personal in nature and can be used by the employee off the job, the payment issue may be left up to labor and management." Examples cited in the memorandum include safety shoes, nonspecialty safety glasses, and cold-weather gear. OSHA makes it clear that, "If shoes and cold-weather gear is subject to contamination of hazardous substances and cannot be safely worn off-site it should be paid for by the employer."

9.5.2 Compliance Requirements

Appendix B of the standard outlines a nonmandatory compliance section regarding hazard assessment and PPE selection. This Appendix outlines general guidelines for identifying, organizing, and analyzing sources of hazards and selection criteria for the appropriate PPE.

As far as we know, OSHA does not plan to issue any compliance directive in the future. It will respond to questions concerning interpretation of the standard. Therefore, without compliance guidance Appendix B will most likely become a significant part of compliance. As history has shown, OSHA is likely to use Appendix B as guidance when applying the standards to a particular situation. Using this nonmandatory section is similar to using the General Duty Clause 5 (a)(1). Employers who fail to follow the nonmandatory section of the Appendix could risk receiving a citation.

Appendix B further describes suggested steps that employers can take when conducting a hazard assessment. According to the Appendix, a survey should include observations of employees and their relation to injury or illness that can occur from work areas where eye, face, head, foot, or hand protection may be necessary to prevent injury from any of the following hazard sources:

- Machinery or processes where any movement of tools, machine elements or particles, or movement of personnel could result in collisions or tripping hazards
- Temperature extremes that could result in burns, eye injury, or ignition of PPE
- Chemical exposures

- Harmful dust that could accumulate or become airborne causing inhalation or physical hazards
- Light radiation sources that could result from operations such as welding, brazing, cutting, furnaces, heat treating, and high intensity lights
- Falling objects or potential for dropping objects
- Sharp objects that might cut the feet or hands
- Rolling or pinching objects that could crush the hands or feet
- Layouts of workplace and location of coworkers
- Electrical hazards

After the hazard assessment has been conducted and the data has been collected, it should be organized in a logical outline that will estimate the potential for employee injury. The organized data will help to decide the type of hazard(s) involved, the level of risk, and the seriousness of potential injury. The appropriate levels of PPE are then selected based on the hazard determination and the availability of PPE. The user should be properly fitted for the specified PPE, and the employer should make sure that it is comfortable to wear. Hazard reassessments should be conducted as necessary based on the introduction of new or revised processes, equipment, and accident experience, to ensure the continued suitability of selection of the proper PPE.

9.5.3 Compliance Issues

OSHA does not specify how the survey data is to be organized or analyzed. Employers should be able to verify that they have conducted an appropriate hazard assessment to identify the level of PPE required to protect the employee from any recognized hazards. The key here is recognized hazards.

A certification document should be developed outlining that the workplace has been evaluated for hazards. It should specify the workplace or areas surveyed and should include the name of the person certifying the evaluation. The contents of a hazard assessment cannot be verified without documentation. Without documentation, the certification could be worthless. So to play it safe, some form of a written certification of the hazard assessment should be retained.

There is no mention that a prior hazard assessment will be acceptable. It is only common sense that OSHA would not expect employers with a previously documented hazard assessment program that meets the new requirements to perform another assessment. Yet there is no way of knowing if this will be acceptable. Employers should use good judgment on what is an effective hazard assessment. Bear in mind, hazard assessment should incorporate any applicable American National Standards

Institute (ANSI) standards for purchases of PPE after July 5, 1994. Therefore, if the prior assessment does not include this, a reevaluation of the population should be conducted.

Employers also have new responsibilities to inspect and remove defective or damaged PPE. It is important that employees are instructed to report defective or damaged equipment.

The new hand protection standard resulted from OSHA's belief that many hand injuries result from not wearing hand protection or wearing protection for the wrong type of hazards. Employers should evaluate and provide hand protection when there are hazards to hands from absorption of harmful substances, severe cuts or lacerations, severe abrasions, punctures, chemical burns, thermal burns, and harmful temperature extremes.

OSHA has warned employers that it will make a special effort to inspect a company's PPE programs to determine whether appropriate equipment was made available and fitted properly to workers. OSHA is particularly interested in female employees. PPE is not always designed properly for women. OSHA plans to interview female employees during inspections to ensure that they are fitted properly.

9.5.4 Employee Training

The training requirements of this standard are more detailed than those of any other OSHA standard. The way it is worded makes it a prime target for OSHA enforcement. The training requirements are written to ensure that employers take the time and effort to train their workers. After the completion of training, each worker should demonstrate an understanding of the training. All employees should be retrained as applicable. The word "applicable" is open to interpretation. When and what is applicable? This is a decision that management should make when training employees.

Documenting training is important to ensure that a company can prove that the requirements have been met. Employers should train each affected employee assigned duties requiring the use of PPE on the following information:

- When PPE is necessary
- What PPE necessary
- How to properly don, doff, adjust, wear, and remove PPE
- Limitations
- The proper care, maintenance, and useful life and disposal of PPE

Employee training is the first step. Before being allowed to work with the designated PPE, employees should demonstrate their understanding of the training requirements and the proper method of using the prescribed PPE.

Now comes the hard part of the training. The employer should verify training through a written process certifying that each employee has received and understood the required instruction. The certification should document the name of the employee trained, the date of the training, and the subject of the training.

OSHA makes it clear in the preamble that the existence of the certification will not preclude a citation if OSHA determines that the employees have not been adequately trained. As a result, employers will need additional records to be able to demonstrate full compliance if there is a disagreement with OSHA. As in the Confined Space Standard, OSHA does not dictate the content or length of the training, or how the employee can demonstrate understanding and competence of the training. It should meet the intent of the standard.

Although there is no requirement that the certification be written, the employer should be able to produce a record of training provided, the methods provided, how an employee was able to demonstrate understanding of the training, and how the employee's ability to use the PPE was confirmed. OSHA may require test results in cases where employee comprehension is in doubt.

Retraining is required when changes in PPE make prior training obsolete. When previously trained employees demonstrate by their behavior that they do not understand when the proper PPE is required or if they are not using it properly, they should be retrained. Employee discipline may also be a controlling factor, and the employers should determine whether the employee's failure to wear the prescribed PPE resulted from lack of understanding of the requirements.

9.5.5 Summary

One major change in the standard is the requirement of a hazard-assessment procedure as outlined in 29 CFR 1910.132 (d). This requirement is meant to ensure that employers make themselves aware of hazards in their work environment. After analyzing hazards and deciding that engineering controls and management practices are not feasible to protect employees, the employer should select and ensure that each affected employee uses the proper types of PPE appropriate for the identified hazard.

A little-known section of the OSHA act applies to the employees (5 (b)): "Each employee shall comply with occupational safety and health standards and all rules, regulations, and orders issued pursuant to this Act that are applicable to his own actions and conduct." Usually the employees do not assume the responsibility for their actions.

Unfortunately, it is up to management to ensure compliance with the standard.

As the awareness of safety and health hazards increases, so does the need to protect workers from these hazards. This need has created an increase in the proper use of PPE. Other factors, such as governmental requirements, worker productivity, and employee morale have stimulated the increased use.

Whether a standard exists or not, companies should realize that operating safely is a responsibility of any corporation and is a part of the cost of doing business. They should realize that operating safely does not rest on the shoulders of government regulation. After all, OSHA standards are minimum performance standards and do not always offer the solution for each situation. It is up to each employer to develop the appropriate solutions to any identified hazards.

Although OSHA regulations and industry standards have begun to address protective clothing and its proper use, the responsibility lies with the buyer for selecting the appropriate type and style of PPE to match the job-specific hazards to protect the worker. When purchasing PPE the construction and quality of the equipment should be kept in mind as well as the regulatory standards that should be met, the comfort and productivity of the worker involved, and the disposability of the equipment after it has been contaminated. The cheapest is not always the best.

9.5.6 Eye and Face Protection

29 CFR 1910.133 discusses eye and face protection for general industry. This standard requires employers to provide the appropriate PPE to protect workers' eyes and faces from situations that could cause injury. According to this standard, prescription lens wearers can be protected by either "safety" glasses and lenses, or protection over the employee's personal eyeglasses. The standard does something unique where it allows side shields that are "detachable" as long as they meet the pertinent requirements. Shaded lenses to protect workers who weld, use cutting torches, braze, or perform other work in which radiation could injure their eyes have specific guidelines for protective lenses in the standard. The standard also refers to ANSI standard Z87.1-1989. This standard is entitled "American National Standard Practice for Occupational and Educational Eye and Face Protection."

Experience tells us that enforcement time spent on the mandatory safety glass program takes less time today than it did in years past. The authors believe that reasons for easier enforcement include:

- Workers are generally more educated than in years past.
- PPE is more widely accepted.

- PPE is generally more stylish.
- There is more effective enforcement for those who wish not to obey rules.
- An aging workforce is more likely to wear prescription lenses full- or part-time when not working.
- PPE is generally more comfortable.

Even though the time spent on requiring full-time eye protection has diminished, eye protection is as important as ever. One field-tested technique that may help to make your eye protection program effective is recognition. You should consider having your supervisors' give a safety meeting on eye protection and then distribute a few pairs of "new style," "extra light weight," "extra heavy duty," or some other kind of innovative safety glasses and have workers test drive them for a month or so. Get some input from these same volunteers to speak up at a meeting in a month or so. Use the input to implement changes in the program.

At some other point, you might have a staff member introduce a new product that is nonfogging or nonscratching. The same technique as discussed above should be utilized. Also, if you have a worker whose sight was saved by safety glasses, recruit that person to use the pair of glasses as a reminder of what occurred. You may consider taking the glasses and putting them in a place where your workers can see them. You may also want to consider honoring compliant workers in public and possibly giving workers an award in appreciation of the occurrence in which safety glasses played a key role.

Get people to talk about safety glasses, eye protection, and safety in general. This is a good thing in general for your safety culture.

9.6 EQUIPMENT LIMITATIONS

The following are some limitations that should be considered when choosing PPE.

Safety Glasses

- Not fitted properly
- Improper use of glasses (prescription vs. nonprescription)
- Dirty, deep scratches, chipped, or pitted lenses will impair vision
- Should choose appropriate glass for application
- Do not protect sufficiently from chemical splashes
- Possible fogging

- Proper glass should be utilized when welding or use of lasers
- Slack, worn-out, sweat-soaked, or twisted headbands do not hold the eyeglasses in proper position

Face Shield

- Does not protect adequately from projectiles
- Scratched or pitted lenses will impair vision
- Does not provide high-impact resistance
- May distort vision

Goggles

- Do not provide high-impact resistance
- May fog, impairing vision

Ear Plugs

- Can interfere with communication if used improperly
- Can introduce contaminants into the ear
- Improper fitting of plug will allow noise to enter ear
- Wrong plug for operation

Ear Muff

- Improper fitting if used with safety glasses, hard hat, etc.
- Poor seal

Safety Shoes

- No protection from punctures
- Do not protect top of the foot
- Do not protect little toe
- Cannot be used in all operations of facility (electrical)
- Improper fitting of shoe
- Poor quality of shoe

Respirators

- Improper fitting of respirator
- Cannot be used with chemicals that do not provide adequate warning properties
- May not be properly selected for the hazard
- May not be properly worn or fitted
- May not be used in oxygen-deficient atmospheres

- Improper fitting due to facial hair, face deformities, eye glasses
- Improper maintenance

Full-Face Respirators

- Facepiece may fog, impairing vision

Half-Face Respirators

- Provide no eye protection
- Provide only partial face protection
- Safety glasses or goggles may interfere

Gloves

- Decrease manual dexterity by adding bulk around fingers (poor fitting, poor grip, stiff)
- Extremely limited for prolonged contact due to permeability
- Not puncture resistant
- Improper glove for wrong job (poor physical properties)
- May be penetrated by many solvents in sufficient degree to be of concern

Hard Hats

- Proper adjustment of helmets is necessary to prevent helmets from falling off
- Paint or cleaning materials may damage the shell and reduce protection by physically weakening it or negating electrical resistance
- Limited use when wearing respirator
- Improper spacing between the webbing and top of head
- Can be affected by sunlight and extreme heat
- Visual signs of dents, cracks, penetration, or any other damage may reduce the degree of safety

9.7 RESPIRATORY PROTECTION

Respirators are discussed in 29 CFR 1910.134. The standard was recently revised and is more comprehensive by far than the older version of the standard. OSHA has placed a renewed emphasis on respiratory protection programs. As we discuss this relatively new standard, we include some figures of respiratory protection currently available.

Figure 9-12 shows a typical half-mask respirator without cartridges attached. One of many types of cartridges that could be used with this

Personal Protective Equipment 133

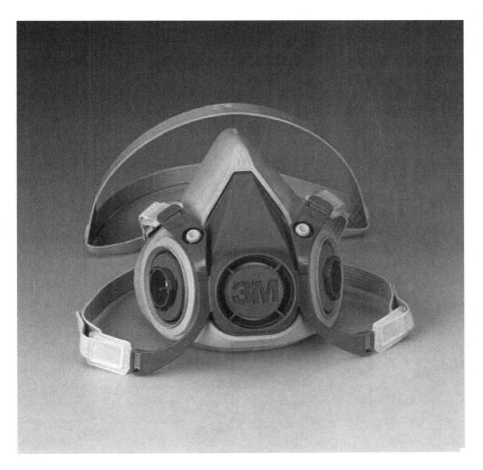

FIGURE 9-12. *A half-mask respirator. Photo: Courtesy 3M*

respirator is shown in Figure 9-13. NOTE: Respiratory protection must be used in accordance with manufacturer's guidance and NIOSH approval. A qualified safety and health professional should be consulted prior to determining respiratory protection needs.

Figure 9-14 shows a typical full-face air purifying respirator (FF APR) with cartridges attached. Figure 9-15 shows what a worker donning modified level C protection (discussed earlier in this chapter) might look like.

Figures 9-16 and 9-17 show an example of a FF APR with a different type of cartridges that can be used. The circular structure below the facepiece window with the cross on it houses the exhalation valve and is not a cartridge holder.

Figure 9-18 shows what the respirator looks like when assembled and ready to use. Figure 9-19 shows the components of a powered air purifying respirator (PAPR). A typical PAPR system uses a small fan to

FIGURE 9-13. *Typical cartridges for an APR. Photo: Courtesy 3M*

blow air through or draw from a filtration system. Filtered air is transported and delivered to the worker via a hose system. This hose system is attached to the fan and to the facepiece. PAPRs have a variety of uses. Some workers feel it is easier to breathe in a PAPR than in an APR. However, the PAPR system is a bit more complicated, cumbersome, and typically weighs more than an APR. Figure 9-20 shows a worker utilizing a PAPR while performing work activities. NOTE: the circular item hanging from the employee's lapel is called a badge. Badges are becoming more commonplace. These badges are used to monitor personal worker exposure to certain chemicals while performing work activities and are not part of the respirator.

Figure 9-21 shows what is commonly referred to as a loose fitting hood. This hood gets a supply of clean air through a hose (not pictured) from the rear of the hood. The air flow is "pushed" past the face of the worker at a relatively rapid rate. The excess air flows out of the hood

FIGURE 9-14. *A full-face APR with cartridges attached. Photo: Courtesy 3M*

FIGURE 9-15. *A worker modeling level C protection. Photo: Courtesy 3M*

FIGURE 9-16. *A view of a full-face APR. Photo: Courtesy 3M*

from the bottom. Thus the worker is breathing only clean air while performing work activities. This type of protection is typically used by workers performing abrasive paint removal, sandblasting, painting, and dusty work.

Figure 9-22 shows some of the main components of an airline system used in level B protection discussed earlier in this chapter. Clean, compressed breathing air is supplied to the worker via compressor, compressed breathing air bottles, or another approved source. The compressed air goes first through a regulator to ensure proper pressure and can then go through a device (controlled by the worker) that can heat or cool the air. A half mask or full-face mask can be used, depending on conditions.

Figure 9-23 shows what a typical worker in Level B protection might look like while working. Quite often the SCBA that this worker is carrying on his back is replaced with an airline type as shown in Figure 9-22. When workers are dressed out in this fashion, the weight of the

FIGURE 9-17. *A view of a typical P100 cartridge. Photo: Courtesy 3M*

SCBA, no matter how light it is, can cause the worker to become tired more quickly than workers using the airline type.

No matter what type of respirator is used, it is of the utmost importance that the revised respiratory standard is adhered to. The revised standard stresses training, documentation, written programs, medical surveillance, fit testing, and a variety of other subjects pertinent to respirators. Of particular interest to the authors is the new approach toward action levels, protection factors, and fit testing. Another important change is OSHA's latest approach on voluntary respirator use. With the new standard in effect, those workers previously considered to be voluntarily wearing respirators should be much better protected.

This new standard applies to all respirator usage in general industry. This includes shipyards, marine terminals, longshoring, and construction workplaces. The standard covers respirator use when they are being worn to protect employees from exposure to air contaminants

FIGURE 9-18. *A view of an assembled FF APR with cartridges. Photo: Courtesy 3M*

above an exposure limit or are otherwise necessary to protect employee health. It also covers workers who are wearing respirators voluntarily for comfort or other reasons.

9.7.1 Permissible Practice

The document restates OSHA's longstanding policy that engineering and work practice controls should be the primary means used to reduce employee exposure to toxic chemicals, and that respirators should only be used if engineering or work practice controls are infeasible or while they are being implemented. Feasible engineering, administrative, or work practice controls should be instituted even though they may not be sufficient to reduce exposure to or below the permissible exposure limit

Personal Protective Equipment 139

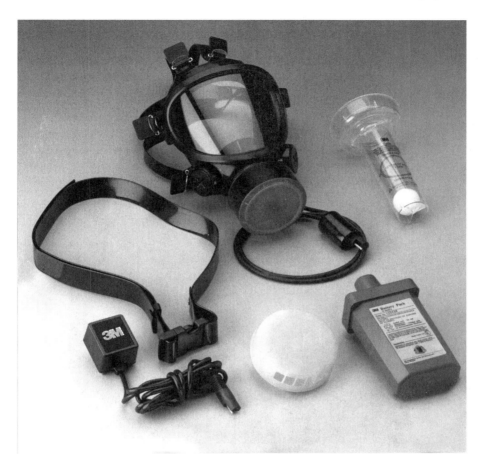

FIGURE 9-19. *Typical components of a powered air purifying respirator. Photo: Courtesy 3M*

(PEL). They should be used in conjunction with respirators when exposure exceeds PELs. The principles discussed for citation guidelines include exceeding a PEL when listed in Table Z of 1910.1000 and for citation under 5 (a)(1) of the act if no specific standard exists.

Whether or not an employer has instituted engineering or work practice controls, the employer's failure to provide respirators when employees are exposed to hazardous levels of air contaminants is citable under 1910.134. The employer should provide the right type of respirator for the substance and level of exposure involved. If the employer provides the wrong kind of respirator, the guidance suggests that a citation could be issued for not providing an appropriate respirator. Where respirators are needed to protect the health of the employees, employers should not only provide respirators but also make sure that employees use them. In cases involving substance-specific standards, the section of

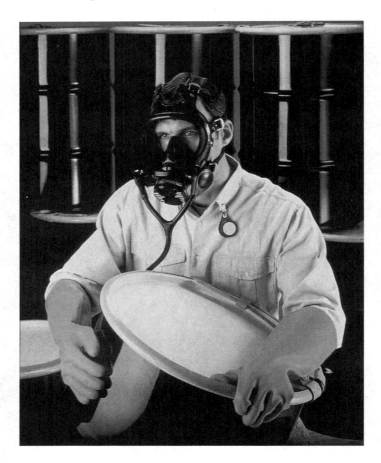

FIGURE 9-20. *This worker is wearing respiratory protection along with a lapel badge which can be used to determine TWA worker exposures. Photo: Courtesy 3M*

the standard requiring respirator use should be cited when employers have not ensured respirator use.

The employer should establish and maintain a respiratory protection program when respirators are required to protect the health of the employee. The program should be in writing and contain all of the elements specified in the standard. If the written program has all of the required elements but the employer has not taken one or more of the actions required, he or she can be cited for each element that has not been met.

9.7.2 Definitions

Some definitions in the proposal were not included in the final standard, and some new definitions were added.

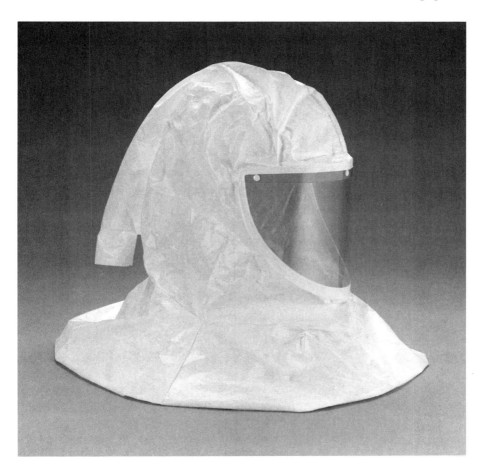

FIGURE 9-21. *A view of a typical hood. Clean air is provided through the hose in the back of the hood and flows over the face of the worker and out the bottom of the hood. Photo: Courtesy 3M*

- **Adequate warning properties:** This was not included in the final standard. OSHA feels the two major warning properties, odor and irritation, are unreliable or otherwise inappropriate to be used as primary indicators of sorbent exhaustion.
- **Assigned protection factor:** This was not included in the standard. However, the latest documentation indicates that OSHA will eventually add APFs into the standard. For now, employers should rely on the best available information when selecting the appropriate respirator.
- **Filtering facepiece:** This could mean a dust mask.
- **HEPA filter:** The efficiency of 99.97 percent used in removing monodispersed particles of 0.3 microns in diameter was considered "HEPA." NIOSH no longer uses this term, but OSHA has retained

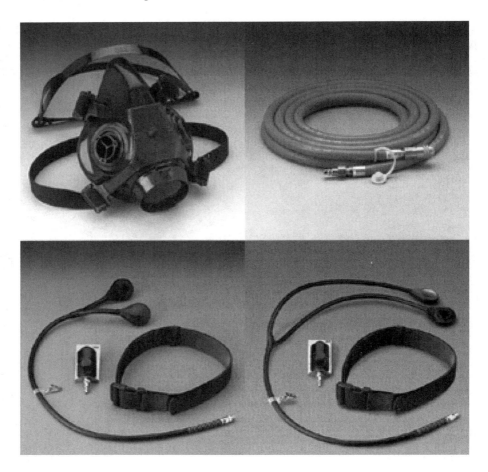

FIGURE 9-22. *Components of airline-type respiratory protection, level B. Photo: Courtesy 3M*

this definition because it is used in many of the existing substance-specific standards. When HEPA filters are required by an OSHA standard, N100, R100, or P100 filters can be used to replace them.

9.7.3 Respiratory Protection Program

A written respiratory protection program is required when necessary to protect the health of the employee from workplace contaminants or when the employer requires the use of respirators. A limited written program is also required when respirators (other than dust masks) are being voluntarily worn by employees. This latest document states: "It is the intent of the standard that the employer would not be required to incur any costs associated with voluntary use of filtering facepieces other

Personal Protective Equipment 143

FIGURE 9-23. *A worker in level B protection. Photo: Courtesy 3M*

than providing a copy of Appendix D to each user." It continues to say, "If employers allow the voluntary use of other than dust masks, medical evaluations and maintenance should be provided at no cost to the employee."

Compliance with the written program can be verified during the walkaround by personal observation and employee interviews. If respirators are required to be worn in the workplace or respirators other than dust masks are worn by voluntary users, a written program is required. An overexposure is not required to cite. Discrepancies between the written program and implemented work practices at the worksite should be cited. Use of a elastomeric or supplied-air respirator, even when voluntary on the part of the employee, will require the employer to include all elements in a written program that will make sure that there is proper use of these respirators so that they do not create a hazard.

9.7.4 Selection of Respiratory and Hazard Evaluation

The employer is required to identify hazardous airborne contaminants that employees may inhale and make a reasonable estimate of employee exposure in determining the appropriate respirator for employees to use. Oxygen-deficient atmospheres and those atmospheres that are not or cannot be estimated should be treated as IDLH environments.

Acceptable means of estimating exposure include:

- Use of objective data (the employer should document the use of objective data as part of their written program)
- Application of mathematical approaches
- Hazards as a result of changes in the workplace

OSHA has warned compliance personnel to use a great deal of professional judgment regarding mathematical approaches. OSHA believes that the results should incorporate reasonable safety factors and be interpreted conservatively.

Appendix A of the revised standard also mentions:

- Experimental methods coming from laboratory-based studies of worst-case testing of simulated workplace conditions.
- Mathematical predictive modeling based on predictive equations.
- Analogous chemical structures. Employers would rely on service life values from other chemicals having analogous chemical structure to the contaminant under evaluation for breakthrough.
- Workplace simulations.

Some general rules of thumb are offered in 1910.134 Appendix A to aid in the assessment.

- If a chemical's boiling point is >70 C and the concentration is less than 200 parts per million (ppm), you can expect a service life of 8 hours at a normal work rate. (This point needs further review.)
- Service life is inversely proportional to work rate.
- Reducing concentration by a factor of ten will increase service life by a factor of five.
- Humidity above 85 percent will reduce service life by 50 percent.

NOTE: The rules of thumb should only be used in concert with one of the other methods of predicting service life for specific contaminants.

9.7.5 Protection against Gases and Vapors on Atmospheres That Are Not IDLH

If a cartridge or canister does not have an end of service life indicator (ESLI), the employer should implement a change schedule based on objective information that will make sure that the cartridges are changed before the end of their service life [4]. The purpose of a change schedule is to establish the time period for replacing respirator cartridges and canisters. This is critical to preventing contamination from respirator breakthrough, and thereby overexposing workers. Data and information relied on to establish the schedule should be included in the respirator program. The new standard prohibits the use of warning properties as the sole basis for determining change schedules. Respirator users should be trained to understand that abnormal odor or irritation is evidence that respirator cartridges need to be replaced.

The change schedule for mixtures is to be based on a reasonable assumption that includes a safety factor. Where the individual components of the mixture have similar breakthrough times (called out as one order of magnitude) the service life of the cartridge should be established assuming the mixture stream behaves as a pure system of the most rapidly migrating components with the shortest breakthrough time. Where the components vary by two orders of magnitude or greater, the service life may be based on the contaminant with the shortest breakthrough time.

9.7.6 Medical Evaluations

Medical evaluation are required for all respirator users except for dust masks, escape only, and others. Employers are required to provide medical evaluation to determine each employee's fitness to wear a respirator. The evaluation should be provided before the initial fit testing and before the respirator is used for the first time.

9.7.6.1 Fit Testing

Fit testing is required for all employees using negative- or positive-pressure tight-fitting respirators, in most cases. A fit test should be performed before the respirator is used in the workplace. It should be repeated at least annually. The new standard contains requirements for fit testing for both qualitative fit testing (QLFT) and quantitative fit testing (QNFT). Employers will still be in compliance with QLFT only when workers are working in atmospheres less than ten times the PEL and using a respirator that achieves a fit factor of 100. For greater concentrations, QNFT should be used. Table 1 of the standard should be consulted to determine acceptable fit testing methods under other scenarios.

9.7.7 Continuing Respirator Effectiveness

The employer is required to address in its written program the type of regular surveillance of the workplace necessary to evaluate the effectiveness of the respirator program. Other items discussed in the standard include:

- Procedures for IDLH atmospheres
- Procedures for interior structural firefighting
- Maintenance and care of respirators
- Respirators available for emergency use
- Breathing air quality and use
- Identification of filters, cartridges, and canisters
- Training and information
- Program evaluation
- Recordkeeping

Consult the complete body of the document for complete information.

9.8 LESSONS LEARNED

Respirators are an important part of worker protection. We discussed a variety of pitfalls and disadvantages earlier in this chapter, for the disadvantages are similar to those found in level A and level B protection. "Selling" respiratory protection is very important. This selling of the program comes through communication, training, and experience. Experienced respirator users know that they work. This faith by workers can be illustrated for workers who work in dusty environments by blowing their noses at the end of the workday. In addition, workers who work with hazardous chemicals that have an odor will notice that the odor is eliminated when wearing the proper respirator.

The downside can be found in overprotection. Some facets of management believe that because respirators are effective, the expanded use of respiratory protection provides somewhat cheap insurance. After all, most workers will typically have their own respirators, so the only additional cost would be cartridges. There is certainly some logic in this type of thinking. However, there is a downside that is usually not taken into consideration.

- Maintenance of the respirator
- Additional stress on workers
- Potential for skin irritation, dermatitis, or skin afflictions from wearing a respirator for extended periods of time or from improper rinsing, cleaning, or maintenance
- Production time lost
- Errors made from the lack of mobility or visibility
- Heat and other stresses

Respirators are great for worker protection, but the administration of any program needs to be carefully implemented and periodically reviewed.

9.9 HEAD PROTECTION

Head protection is discussed in 1910.135. The employer should make sure that affected workers wear head protection when working in areas where there is a potential for injury to the head from falling objects or to reduce electrical shock hazard. ANSI standards Z89.1-1986 and Z89.1-1969 are incorporated by reference.

9.10 FOOT AND HAND PROTECTION

Foot protection is mentioned in 1910.136. ANSI standard Z41-1991 and Z41.1-1967 are incorporated by reference. Hand protection is covered in 1910.138. The employers should make sure that workers are wearing appropriate hand protection.

9.10.1 Lessons Learned

Foot and hand protection are basic concepts. Unfortunately, in studying numerous OSHA logs, there appear to be many injuries that might have been prevented with adequate foot and hand protection. Hand protection holds particular interest. It would appear that for almost any work task, some type of glove or hand protection would be in order. This might mean tight fitting latex gloves, cotton or leather worker gloves, chemical resistance gloves, wire mesh gloves, electrical lineman gloves, or many others. If the injuries that your workers are suffering occur to their hands or feet due to a lack of protection, a serious look at this part of your program would be in order.

REFERENCES

1. 29 CFR 1910.132 "OSHA Personal Protective Equipment Standard."
2. 29 CFR 1910.120 "OSHA HAZWOPER Standard."
3. *Handbook for Occupational Health and Safety During Hazardous Waste Activities.* Office of Environmental, Safety and Health Office of Environmental Management, 1996, pp. 7-13–7-15.
4. 29 CFR 1910.134 "OSHA Respiratory Protection."

Chapter 10

Decontamination Activities

Decontamination is the process of removing or neutralizing a chemical, radiological, biological, or mixed waste (or all contaminants) that collects on workers, personnel, or equipment while work is being performed [1]. Contamination control is a critical element to consider when trying to protect the workers, the public, and the environment when working with hazardous materials [2]. Worker and equipment decontamination is a major concern when dealing with hazardous materials. It is important that PMs understand the importance of decontamination and contamination control when planning these activities.

Anything that enters an exclusion zone should be evaluated as to its potential contamination. If not removed properly, these contaminants may permeate PPE, tools, instruments, and other equipment [2]. In addition, this potential contamination can be transferred to the clean zones if it is not controlled. If contamination does get into clean zones, anyone may be able to take the contamination home with them to affect themselves, family members, and the general population.

Effective planning is again the key. We need to keep in mind that proper decontamination can be costly, but improper decontamination can be even more costly. One large but variable cost is the time it takes workers to decontaminate. The time that decontamination will take should be estimated and incorporated into the budget. In addition, contamination control and decontamination strategies and procedures should be outlined in the safety plan, communicated to workers, and implemented before any worker enters any area where there is a potential to become contaminated.

The safety plan should specify the level of decontamination necessary for specific site activities. Appropriate procedures should be developed and implemented to help minimize contamination, to prevent the spread, and to decontaminate workers and equipment when they exit any contaminated area [1].

As a general rule, contamination control procedures depend on

- The type and source of contaminants
- The level of contamination

- The severity of the hazard exposure
- The evaluation of worksite hazards
- The job tasks to be performed

If the source of contamination is an extremely hazardous or dangerous material and the task at hand requires that workers come in contact with this dangerous material, plan on extra time for decontamination. On the other hand, if the hazardous material borders on nuisance levels, and can easily and readily be removed, or workers use only disposable clothing, decontamination should take much less time.

Contamination control processes specified in the safety plan must be periodically evaluated for effectiveness and modified to correct any deficiencies noted and address changing conditions and activities [3]. This periodic evaluation for effectiveness has also been the subject of much debate. You might ask, "What exactly is a periodic evaluation?" Typically, persons managing the site would like hard and fast rules for periodic evaluations. However, the HAZWOPER standard leaves this determination up to the employer. As you read through this chapter, you will see why no specific time line or procedure is set for the periodic evaluations.

10.1 DECONTAMINATION STRATEGY

Decontamination protocols should be designed to remove hazardous substances from workers, PPE, and other equipment exiting contaminated areas. A protocol could be as simple as doffing PPE and placing it into appropriate containers for disposal or decontamination.

The authors believe that increased use of disposable clothing has made decontamination easier in most cases. Besides disposable clothing, disposable respirators may also be an advantage in certain instances. Although there is typically a higher initial cost with disposables in general, there is also substantial benefit—keep in mind that in addition to the initial cost you should take into consideration disposal cost. If you are disposing of your disposable clothing as hazardous waste (which is sometimes done because of the mere convenience and easy availability), the costs for disposal can be quite high. However, if you sample your disposal clothing prior to disposal to determine the type of waste, you need to factor in sampling time and laboratory costs.

10.1.1 Time Savings in Decontamination

A typical decontamination program should contain the following:

- Documentation of the hazards, and how those hazards are anticipated to be removed

- Specific decontamination methods that will be used, including specific instruments such as brushes that will be used along with detergents or fluids for neutralization
- Testing for decontamination effectiveness, which might include analysis of the decontamination fluid along with visual inspections of personnel, equipment, and fluids
- Location and configuration of the decontamination area
- Emergency decontamination procedures
- Identification of decontamination hazards
- Protection of decontamination workers
- Disposal methods, equipment decontamination
- Sanitation
- Waste minimization

Each protocol specifies what personal hygiene practices (from hand washing through extensive decontamination showering) are necessary. This should depend on the type and degree of the hazard. Various methods of cleaning, neutralizing, or removing contaminants should be evaluated for use. Decisions concerning decontamination approaches should be based on the extent of site-specific hazards and activities. If not already specified in the safety plan, all aspects of the decontamination approach and program should be documented in a decontamination plan. This plan should address the following elements:

- The number, location, and layout of decontamination stations
- Decontamination equipment that may be needed (brushes, buckets, etc.)
- Appropriate decontamination methods (high-pressure wash)
- Procedures to prevent contamination of clean areas (appropriate barriers, plastic sheeting, etc.)
- Methods and procedures to minimize worker contact with contaminants during removal of PPE
- Methods for disposing of clothing and equipment that are not completely decontaminated
- Incompatible wastes requiring separate decontamination stations (metal drum vs. plastic drum)
- The target level of decontamination

The plan should also address standard operating procedures (SOPs) for site operations to help minimize contact with hazardous materials. Some examples of typical SOPs may include:

- Work practices that minimize contact with hazardous substances.
- Use of remote sampling, handling, and container opening techniques. This can be achieved with robots, or, more commonly, by using

long-armed back hoes or heavy equipment with grapplers or a bronze spike.
- Protection of monitoring and sampling instruments by covering them with plastic or plastic bags (openings can be made in the bags for sample ports and sensors that are required to physically contact worksite materials).
- Wearing disposable outer garments and using other disposable equipment as applicable.
- Covering equipment and tools with a strippable coating that can be removed during decontamination.
- Encasing the source of contaminants with, for example, plastic sheeting or overpacks [2].

10.2 ACCEPTABLE DECONTAMINATION METHODS

To prevent the further generation of mixed wastes, decontamination methods should be chosen carefully and implemented to be part of the overall solutions and not part of the cleanup [2].

10.2.1 Contact Time

Contaminants can be deposited on the surface of or can permeate PPE and other equipment. The longer a contaminant stays in contact with an object, the greater the probability and extent of permeation. Minimizing contact time is one of the most important objectives of a decontamination program. This is why contact time with hazardous material should be taken into consideration when considering different methodologies in the actual performance of the task itself.

Most surface contamination is detected and removed by accepted decontamination practices. If a contaminant has permeated the PPE (i.e., the fabric of coveralls), it may be difficult to detect and remove. When contaminants are allowed to remain in contact with materials for an extended period, those materials are prone to permeation or degradation [2].

10.2.2 Concentration

As concentrations of contaminants increase, the potential for permeation of PPE increase. The chemical and physical compatibility of decontamination solutions and methods with selected PPE should be determined before use [2].

10.2.3 Temperature

Temperature increases generally increase the contaminant permeation rate [2]. However, temperatures between 40° and 90° F usually do not have a significant effect.

10.2.4 Chemical Characteristics

Permeation rates are dependent on the chemical makeup of the contamination. This includes the size of the contaminant (how large or small the molecule or particle is) and on the pore size of the protective material (for instance, impermeable rubber suits, tyveks, or cotton coveralls). Chemical characteristics (i.e., polarity, vapor pressure, pH) of both the contaminant and the protective material also determine permeability. Keep in mind that gases, vapors, and low-viscosity liquids tend to permeate more readily than high-viscosity liquids or solids [2].

10.2.5 Decontamination by Physical Means

Some contaminants encountered are removed by physical means (i.e., washing, brushing, scraping, using sticky tape, rinsing, heating) that dislodge or displace the contaminant. Caution should be taken when selecting physical methods involving high pressure or heat because these methods can produce aerosols, penetration, cut, burns, or hazards associated with the equipment. In addition, weather conditions should be considered when choosing physical decontamination methods [2].

Contaminants that can be physically removed fall into four major categories.

10.2.5.1 Loose Contaminants

Dusts and aerosols that cling to equipment and workers or become trapped in small openings (i.e., in the weave of fabrics, behind bulkheads, etc.) can be removed with sticky tape, water, or a liquid rinse. Removal of electrostatically attached material is increased by coating clothing or equipment with antistatic solutions. Chemicals can be complexed (i.e., metals precipitation) and removed using specially designed vacuums equipped with HEPA filters and other system controls; asbestos fibers can be removed using similar devices. In some cases, lead, asbestos, and elemental mercury can be removed using special vacuums.

10.2.5.2 Adhering Contaminants

Removal is often enhanced through methods such as solidification, freezing (i.e., ice or dry ice), adsorption or absorption (i.e., powdered lime or kitty litter), or melting with a low-energy heat source (i.e., hair dryer or heat lamp).

10.2.5.3 Adsorbed or Permeated Contaminants

In some cases the contaminant cannot be removed. In this case, the PPE tools, instruments, or other equipment should be discarded (as hazardous waste, if necessary). Care in selecting PPE and in applying contamination prevention and control measures, along with timely and appropriate decontamination measures, will often prevent this situation. In particular, if shovels, scrapers or other implements should be used within the exclusion zone, specify a material other than wood or other materials that are porous. The same goes for ladder choices, or other items that find their way into the exclusion zone.

10.2.5.4 Volatile Liquids

Volatile liquid contaminants can be removed from PPE or equipment by evaporation followed by a water rinse. Evaporation of volatile liquids can be enhanced by using steam jets. With any evaporation or vaporization process, care should be taken to prevent worker inhalation of vaporized chemicals. And, of course, the physical hazards of steam need to be taken into consideration along with protection needed to control any splatter of liquid or debris. Regulations pertinent to air emissions must also be taken into consideration.

10.3 USING SOLUTIONS, CHEMICALS, AND OTHER MATERIALS

Physical removal of hazardous substances should always be followed by washing or rinsing. Steam (for equipment) or hot water with detergent is the preferred decontamination method. In some cases, it may be necessary to use a special solution or combination of solutions to decontaminate thoroughly. The SSHO should consult with the appropriate engineers, chemists, toxicologists, or other individuals for selection of the safest and most effective decontamination solutions for the specific contaminants. Selection is influenced by hazards posed by the decontamination method, effectiveness of the decontamination method, ease of implementation, availability, and cost.

Cleaning solutions normally use one or more of the following methods:

- **Dissolving contaminants:** Chemical removal of surface contaminants can be accomplished by dissolving them in a solvent. It is important to make sure that the solvent is chemically compatible with the equipment being cleaned. This is particularly important when decontaminating PPE constructed of organic materials that could be damaged or dissolved by organic solvents. Care should be taken in selecting, using, and disposing of any organic solvents that may be flammable or potentially toxic. Organic solvents include alcohol, ethers, ketones, aromatics, straight-chain alkanes, and common petroleum products. Halogenated solvents generally are incompatible with PPE and are toxic. They should be used for decontamination only in extreme cases where other cleaning agents will not remove the contaminant. Although many solvents are available, the waste stream created from using many solvents makes this scenario less desirable.
- **Surfactants:** Surfactants augment physical cleaning methods by reducing adhesion forces between contaminants and the surface being cleaned, and by preventing redeposition of the contaminants. Household detergents are among the most common surfactants. Some detergents can be used with organic solvents to improve the dissolving and dispersal of contaminants into the solvent. Biodegradable solvents have been the detergent of choice in recent years.
- **Solidification:** Solidifying liquid or gel contaminants can enhance the physical removal. The mechanisms of solidification can be described as follows:
  - Moisture removal through the use of adsorbents such as ground clay, flyash, or powdered lime
  - Chemical reactions via polymerization catalysts and chemical reagents
  - Freezing using ice water

  We need to keep in mind the disposal costs in all of the mechanisms for solidification. With the first method, keep in mind that free liquids are typically not allowed in most disposal scenarios. And adding too much adsorbent can substantially add to disposal costs. Make this point clear to your field people. As far as using polymerization catalysts and chemical reagents, keep in mind disposal costs. Ensure that you are cognizant of disposal costs of spent catalyst prior to using this scenario. As far as freezing is concerned, consider the cost to keep the contaminants frozen and what the downsides are. The downsides besides cost include measures in case of power failure and use of freezing equipment after wastes have been disposed.
- **Rinsing:** Rinsing removes contaminants through dilution, physical attraction, and solubilization. Multiple rinses with clean solutions remove more contaminants than a single rinse with the same volume

of solution. Continuous rinsing with large volumes of clean solutions will remove even more contaminants than multiple rinsings with a lesser total volume. Keep in mind the disposal costs of the cleaning solution used for rinsing.
- **Disinfection/sterilization:** Chemical disinfectants are a more practical means of inactivating infectious agents when compared to sterilization. Standard sterilization techniques are generally impractical for large equipment and for nondisposable PPE. For this reason, disposable PPE is recommended for use with infectious agents [1].

10.4 DETERMINING DECONTAMINATION EFFECTIVENESS

The effectiveness of the decontamination method used should be assessed at the beginning of a project and periodically during the remediation period. If contaminants are not being removed or are penetrating protective clothing, the decontamination program should be revised. There are several useful methods in assessing the effectiveness of decontamination.

10.4.1 Visual Observation

Visual observation involves the use of natural and ultraviolet light. In natural light, discolorations, stains, corrosive effects, visible dirt, or alterations in clothing fabric may indicate that contaminants have not been removed. In ultraviolet light, certain contaminants (i.e., polycyclic aromatic hydrocarbons, common in many refined oils and solvent wastes) fluoresce and can be detected visually. Ultraviolet light can be used to observe contamination of skin, clothing, and equipment. A qualified health professional should be consulted prior to the use of this technique, since certain areas of the skin may fluoresce naturally and introduce uncertainty into the test. In addition, use of ultraviolet light can increase the risk of skin cancer and eye damage.

10.4.2 Wipe Sampling

Wipe sampling involves swiping a dry or wet (use of a solvent or other liquid besides water in commonplace) cloth, glass fiber filter paper, or swab over the surface of a potentially contaminated object and performing a laboratory analysis. Both the inner and outer surfaces of PPE should be tested to check for permeation. Skin can also be tested using this wipe sampling [1].

10.5 CLEANING SOLUTION ANALYSIS

Analysis of contaminants left in cleaning (or final rinse) solutions may indicate that additional cleaning and rinsing are necessary [1].

10.5.1 Permeation Testing

Testing for the presence of permeated chemical contaminants requires that pieces (in the case of contaminated soil this would be typically those that visually indicate contamination) of the protective garments be sent to a laboratory for analysis.

10.6 DEFINING DECONTAMINATION AREAS

Decontamination is conducted in the contamination reduction corridor (CRC), in a defined contamination reduction zone (CRZ), or in a radiological buffer area. Decontamination equipment, processes, and procedures vary, as do contamination reduction zones and corridors, depending on the presence of specific hazards and the size and complexity of the worksite. Modifications to the location and configuration of the decontamination area will likely be required to accommodate changing conditions at the worksite [1]. Changing conditions may be the wind but may also include work activities, out-of-the ordinary contamination discovery, logistical considerations, and others.

10.7 EMERGENCY DECONTAMINATION PROCEDURES

The project team plans for both routine and emergency decontamination and documents the plans in the safety plan. To prevent the possibility of decontamination causing serious health effects or aggravating existing illnesses or injuries, methods should be established for decontaminating workers with medical problems or injuries. When PPE is grossly contaminated, it is possible that contaminants can be transferred to either emergency medical personnel or the wearer. Unless severe medical problems have occurred simultaneously with gross contamination events, PPE is quickly washed off and carefully removed.

A worker who is suspected of having inhaled harmful levels of contaminants should be removed from the area immediately and receive appropriate first-aid treatment while he or she is waiting for treatment by a physician.

If the contaminant is on the skin or in the eyes, immediate measures should be taken to remove it and counteract its effects. First-aid treatment usually involves flooding the affected area with clean water for at least 15 minutes. For a few chemicals, water may cause more serious problems [1]. The safety plan should anticipate and contain procedures for dealing with such possibilities.

Lifesaving care should begin immediately without considering decontamination. Difficulty in breathing, cardiac arrest, arrhythmia, heatstroke, and severe bleeding should be treated as quickly as possible. In addressing life-threatening circumstances, the following actions are to be considered:

- Outer garments and PPE may be removed depending on injury, weather conditions, delays, interference with treatment, or aggravation of the problem.
- Respirators and backpack assemblies should be removed.
- Fully encapsulating suits or chemical-resistant clothing can be cut away.
- If removal of contaminated garments will cause further injury, the individual should be wrapped in plastic, rubber, or blankets to prevent contamination of medical personnel and equipment.
- Contaminated garments should be removed at a medical facility, and carefully handled and contained to prevent or minimize cross-contamination.
- No attempt should be made to wash or rinse the victim at the worksite unless the individual is known to be contaminated with an extremely toxic or corrosive material that could cause further severe injury or loss of life.
- For minor medical problems or injuries, normal decontamination procedures are to be followed [2].

10.8 IDENTIFICATION OF DECONTAMINATION HAZARDS

Decontamination of PPE reduces exposures and protects worker safety and health. However, physical and chemical decontamination methods may themselves be hazardous. Methods that permeate, degrade, damage, or reduce PPE effectiveness should be avoided. PPE, sampling instruments, tools, and other equipment are usually decontaminated by scrubbing with solutions of detergent and water, using soft-bristle brushes, followed by rinsing with water. Though this process may not remove all contaminants completely (or in a few cases, contaminants may react with water), it is safer than using harsh chemicals.

Potential decontamination hazards include, but are not limited to, the following:

- Incompatibility between decontaminating agents and contaminants
- Incompatibility between decontaminating agents and clothing or equipment being decontaminated
- Potential effects of inclement weather (i.e., using wet procedures during cold weather can cause both operational and maintenance problems)
- Potential effects of hazards on worker S&H (i.e., vapors from chemical decontamination solutions may be hazardous on inhalation or contact with skin, or may be flammable)
- Generation of airborne contaminants from improper use of equipment (i.e., jet sprayers, vacuum cleaners) [2]

10.9 PROTECTION OF DECONTAMINATION WORKERS

A JHA should be conducted and hazards associated with the decontamination process should be identified to determine the appropriate types of PPE for decontamination workers. This information should be incorporated in the safety plan.

For many operations, workers are assigned to assist in conducting decontamination of workers wearing Level A or B PPE during the decontamination process. Decontamination workers stationed at the front end of the decontamination line may require more protection from chemical and radiological contaminants than decontamination workers assigned to the latter stages of decontamination.

In some cases, decontamination workers wear the same level of PPE as workers entering the controlled area or exclusion zone. In others, decontamination workers are sufficiently protected by wearing a lower level of PPE. In many instances, level D protection is not acceptable in the CRZ. In addition, all decontamination workers should be decontaminated before entering the support zone. Appropriate equipment and clothing for protecting decontamination workers should be planned by the project team [2].

10.10 DISPOSAL METHODS

Before other operations begin, all materials used in the decontamination of workers and equipment should be disposed of properly. Materials used for decontamination are regarded as hazardous, radioactive, or mixed waste until adequately evaluated and an accurate determination is made. Buckets, brushes, clothing, tools, and other contaminated equipment are collected and labeled appropriately. Yellow plastic wrapping material should be used for packaging radioactively contaminated material. Yellow plastic sheets or bags should not be used for nonradiological purposes. Care should be taken to avoid placing waste streams of incompatible contaminants together in the same container and to emphasize waste minimization methods whenever possible [2].

10.11 EQUIPMENT DECONTAMINATION

Although contamination reduction or total avoidance of contamination is preferred, typically, some equipment used in remedial actions or sampling becomes contaminated. These items are either properly decontaminated before being removed from the site or, in the case of drilling tools, thoroughly cleaned before the next use. Disposable plastic tarpaulins can be used to minimize the need for subsequent cleaning. Particular care should be given to such elements as tracks, tires, shovels, grapples, and scoops that come into direct contact with contaminants.

A thorough inspection of equipment, which may include frisking or a wipe test, helps determine the duration of and methodology selected for decontamination. All equipment parts should be considered highly contaminated, removed, and replaced before the equipment leaves the worksite. Porous items (i.e., wooden truck beds, cloth hoses, wooden handles) usually cannot be thoroughly cleaned and should be discarded (as hazardous waste if necessary).

Decontamination of vehicles and large pieces of equipment (i.e., pumps) is typically conducted on a wash-pad constructed so that cleaning solutions and wash-water are recycled or collected for later disposal. Similarly, equipment being dry-brushed or vacuumed with specially filtered vacuums (HEPA filters) is placed on a nonporous pad to facilitate containment and waste collection.

Decontamination starts with the simplest methods likely to be effective (i.e., a general wet spraying to remove most of the contamination followed by scrubbing more difficult areas). By following procedures such as these, workers are able to minimize unnecessary contact with contaminated equipment. Steam cleaning and pressure spraying using water mixed with a general-purpose, low-sudsing soap or detergent to improve wetting is the preferred method for wet decontamination. Scrubbing with disposable or easily decontaminated brushes may be necessary to loosen materials. In most instances, hot water is more effective than cold. Flushing should be done under high pressure, taking care not to damage dials, gauges, wires, or hoses. Power spraying is often more effective for such items as shovels, loaders, and scoops. Dry removal of contaminants can be accomplished through brushing, vacuum cleaning, vacuum blasting, and sandblasting. Vacuum cleaning with HEPA units can provide an adequate control mechanism for fugitive emissions [2].

10.12 SANITATION

The concepts of sanitation and decontamination are sometimes confused. Sanitation is the promotion of general public health by controlling sewage, protecting the cleanliness of drinking water, and promoting personal hygiene. Decontamination involves eliminating or deactivating either radiological or nonradiological contaminants and preventing the migration of hazardous constituents outside the worksite boundaries.

Many hazardous waste activity worksites are temporary and are established at remote locations with limited sanitation facilities. Decontamination is conducted either in the contamination reduction zone or the radiological buffer zone, whereas sanitation functions are performed either in the support zone or outside the boundaries of the hazardous waste activities worksite after decontamination has been completed.

For jobs lasting 6 months or longer, showers and two-stage changerooms are provided in accordance with 29 CFR 1910.141 (d). When working with asbestos or lead, a five-stage decontamination may be the method of choice. These five stages include:

- **A dirty room:** A dirty room is where workers remove contaminated clothing. It should be located directly next to the exclusion zone. Keep in mind that the gross decontamination needs to be performed within the exclusion zone. As workers are exiting the exclusion zone, after they have performed gross decontamination they will proceed directly into the dirty room. It is important to follow through with training and inspections to ensure workers do not get the dirty room grossly contaminated. Dirty room lockers are recommended. Each worker should be assigned his or her own locker.
- **Dirty air lock:** Once the workers have deposited their "dirty" clothing into laundry containers or dirty clothes lockers, they proceed through the dirty air lock with their respirators still on. The dirty air lock makes sure that dirt from the dirty room does not migrate into the shower area or further.
- **The shower area:** The shower area may have one or more showers where the worker, without clothes, takes a full body shower. It is in the shower that the worker would remove the respirator and place cartridges in a proper receptacle for disposal. The worker's respirator should be cleaned while in the shower area.
- **Clean air lock:** After exiting the shower, the worker proceeds through a clean air lock area toward the clean room. This area has been used for hanging up respirators after cleaning to air dry. From the clean air lock the worker proceeds into the clean room.
- **Clean room:** The clean room is an area where showered workers dry off, dress up in their street clothes, and exit the decontamination area. The clean room should have a clean locker for every worker. The

cleanliness of the clean room and the lockers is very important. We have found it effective to assign remediation workers to clean this area. These workers can help to police the area so that high levels of cleanliness can be attained.

The order used to clean these areas may be intuitive, but for clarity's sake we should keep in mind the following principle. Cleanup should be performed in the cleanest areas first. For the five-stage decontamination area mentioned, cleaning should start in the clean room. In fact, cleaning should start in the cleanest area of the clean room, and then address each area in order of cleanliness. The last area to be cleaned will be the dirty room. If workers will be performing cleanup activities, they should be properly trained and qualified. If an outside service is utilized, those workers should also be trained and qualified [4].

Keep in mind that the decontamination area should be designed to accommodate both genders, as applicable. It is important that workers feel confident that the decontamination area and program are effective. Should workers express discomfort in the decontamination area due to gender concerns, it might be advantageous to install two decontamination areas. If workers are uncomfortable when showering or decontaminating themselves, they may not perform adequate decon, and end up spreading contamination.

Access to emergency showers and eyewashes is part of the site-specific emergency response and medical first-aid programs, and is unrelated to sanitation or decontamination. Requirements for the availability and location of emergency showers and eyewashes are specified under 29 CFR 1910.151.

The hazardous waste standard requires employers to make certain that when showers are a necessary step in the decontamination process, "their employees shower at the end of their work shift and when leaving the hazardous waste site." Sanitation-related showers (unlike decontamination showers) are understood to be voluntary. Decontamination and emergency showers should be located close to the worksite. Sanitary showers may be located at some distance from the worksite. A statement in the safety plan encouraging good personal hygiene and daily showers is a good idea. In addition, workers should be encouraged to shower daily even if no shower is available at the worksite.

10.13 WASTE MINIMIZATION

Although waste minimization is not specifically addressed in the HAZWOPER standard, it does represent a management practice that supports worker and equipment decontamination. Waste minimization practices help to protect the environment and decrease project costs (through reduced storage and transportation requirements), reduce worker exposures, and decrease the overall infrastructure required for decontamination [2].

REFERENCES

1. *Occupational Safety and Health Guidance Manual for Hazardous Waste Site Activities.* Prepared by National Institute for Occupational Safety and Health (NIOSH), Occupational Safety and Health Administration (OSHA), U.S. Coast Guard (USCG), U.S. Environmental Protection Agency (EPA), October 1985, pp. 9-2, 10-2–10-7.
2. *Handbook for Occupational Health and Safety During Hazardous Waste Activities.* Office of Environmental, Safety and Health Office of Environmental Management, 1996, pp. 8-1–8-9.
3. 29 CFR 1910.120 "OSHA HAZWOPER Standard."
4. *Model Curriculum for Training Asbestos Abatement Contractors and Supervisors.* Safety Health and Ergonomics Branch Electro-Optics, Environment, and Materials Laboratory, U.S. Environmental Protection Agency Cooperative Agreement No. CX 820760-010-0, pp. X-6, X-7, X-8.

Chapter 11

Emergency Preparedness and Response

Anytime work is being performed with hazardous substances there is a possibility that an emergency may occur. Emergencies happen quickly and unexpectedly and require immediate response. At a hazardous waste site, an emergency may be as limited as a worker experiencing heat stress, or as major as a fire, explosion, or release [1]. HAZWOPER has established requirements to provide protection for employees who are involved in hazardous waste operations and emergency response. 29 CFR 1910.120 (a)(q) applies to employers who have workers who are expected to perform emergency response to releases, or potential releases, of hazardous substances, regardless of location.

Unless employers can demonstrate that their operation does not involve employee exposure or the reasonable possibility for employee exposure to safety or health hazards, they should comply with the standard. To determine if your particular situation is covered by the emergency response provisions of the standard, examine the tasks of your workers to determine if they will be assigned a role or function as part of a response to a release of hazardous waste [2].

We mention the hazardous waste standard due to the specific requirements of this standard. However, should your operation involve hazardous materials, the same basic principles apply. Those principles, simply stated, are that workers should be properly trained, qualified, and prepared to perform their work. If their work is responding to an emergency situation or release, the worker should be able to do so without becoming injured. It does not matter if your site is a hazardous waste site or not; workers should be adequately prepared to perform expected work.

Workers on a hazardous waste site are not allowed to participate in any emergency response activity unless they are in compliance with the requirements of 29 CFR 1910.120 (e.g., responders to the scene would have to be covered, but operators such as truck drivers would not have to be covered unless they become actively involved in the response action). 29 CFR 1910.120 (q) applies to all organizations that respond to uncontrolled releases of hazardous wastes or substances. Sites where emergency response operations take place, and that do not fall into any of the other categories listed in paragraphs (a)(1)(i) through (a)(1)(iv), must comply at least with the requirements of paragraph (q) of the hazardous waste standard. In contrast, sites that have the possibility for hazardous waste activities under paragraphs (a)(1)(i) through (a)(1)(iv) must comply with multiple paragraphs of the hazardous waste standard.

Sites that do not establish their own emergency response capabilities and elect to evacuate all employees should develop an emergency action plan (EAP) (an evacuation plan) in accordance with 29 CFR 1910.38. Even if you plan to evacuate all personnel, you need to plan for an emergency [2]. If you do not have in-house emergency response experts at your disposal, you should consider looking outside of your organization. There may be more resources available than you realize, but a good first step would be contacting your local fire department.

The local fire department may have all of the resources that your site would need in handling the worst possible site emergency. At the other extreme, the fire department may be not be equipped properly, may be poorly qualified, or may be unable to respond quickly or adequately to a site emergency without adequate assistance from other sources. No matter what the level of competence of the local fire department, they typically know how emergency situations should be handled and know where to look to get the assistance needed should an emergency occur. If there is a chemical plant or factory nearby, there is a good possibility that the local fire department has an agreement with the factory to use their emergency equipment should the need arise.

A good relationship with local fire departments, local police departments, and city officials is encouraged. We recommend working closely with all branches of government, especially fire and police departments. We believe that a formal meeting should take place between site management and local entities. This meeting should be documented in the form of minutes. During these meetings you should define the role of off-site responders, address the need for off-site resources to support pre-incident planning, and provide for the availability of adequate off-site response capabilities in an emergency. These meeting minutes should be forwarded to all parties in attendance and used as a vehicle for future meetings. In a formal atmosphere you might call these items memoranda of understanding (MOU) or memoranda of agreement (MOA) with the local fire department or hazardous materials (HAZMAT) response team [2].

11.1 EMERGENCY RESPONSE

The HAZWOPER standard says that an emergency exists when an incident occurs that results in, or is likely to result in, an uncontrolled hazardous waste or substance release. This emergency would cause a potential health or safety hazard that cannot be mitigated, or "fixed" by personnel in the immediate work area where the release occurs [2].

In the case of fire, the term "incipient" is often used. This means that the fire can easily be controlled and extinguished by the discoverer without the likelihood that the discoverer of the fire would be injured during the response. An example of this might be the discovery of a paper fire in a waste paper basket. The discoverer who believes that he or she can extinguish this fire without likelihood of injury can do so without being HAZWOPER Emergency Responder trained. However, they might still need hazard communication training and fire extinguisher training. An incipient event would not be an event in which trained responders from outside the immediate work area (which may include other site or facility response personnel, mutual aid groups, or the local fire department or HAZMAT team) are relied on for response.

In 29 CFR 1910.120 (a)(3) it is stated that responses to incidental releases of hazardous substances where the substance can be absorbed, neutralized, or otherwise controlled at the time of release by employees in the immediate release area, or by maintenance personnel, are not considered to be emergency responses in the scope of the standard (HAZWOPER). The term incidental is the key term. Workers need to be trained as to what type of situations would be considered incidental. In general, if the employees' actions to clean or control the release do not and likely would not put them in jeopardy (from a safety and health viewpoint), the act would be considered incidental.

The term incidental is analogous to the term incipient in the example of the waste basket fire. Whether using the term incidental or incipient, the principle is the same. The questions we need to answer are:

- Is the worker trained and qualified to perform this duty?
- What is the potential for the emergency to get out of hand?
- What is the potential for the situation to change from incipient or incidental to a more serious situation?
- What is the potential for worker injury when performing this duty?

These questions need to be considered for every emergency situation. And the workers need to be able to make judgment calls in a very short time frame. Whether to attempt to handle an emergency or walk away and report it becomes very situation dependent. Keep in mind that situations with real-life potentials should be discussed during training.

In addition, responses to releases of hazardous substances where there is no potential health or safety hazard (i.e., fire, explosion, or chemical exposure) are not considered to be emergency responses. Keep in mind that qualified personnel who are trained to clean up incidental releases under the Hazard Communication Standard (HAZCOM) are not considered emergency responders.

Consider the following scenario. A small quantity of sodium hypochlorite is spilled in a wastewater treatment process. The maintenance engineer who is normally assigned to the immediate work area mops it up. This situation is not considered an emergency response because the situation as described would be considered incidental. The engineer needs to be qualified to do this task but does not have to be trained in emergency response. In this example, the worker would be expected to understand the hazards associated with sodium hypochlorite through previous training. This training would include HAZCOM and other training [2].

Let's make a few adjustments to this scenario and see how this changes the situation. Let's keep everything the same except for the leak. Let's assume that in the new situation the leak is large and uncontrolled. These two words, large and uncontrolled, would likely change the status of this situation. Obviously, there will be judgment calls for both the terms large and uncontrolled. Again, training is the key.

For extremely dangerous substances (not necessarily sodium hypochlorite), possibly a quart, pint, or even less could be considered a large quantity. Yet for a constituent that is considered a bit more dangerous than nuisance, the large quantity might be many gallons or cubic yards.

Let's go back and adjust the initial scenario again. This time, let's keep the quantity small but change personnel. In the original scenario a maintenance engineer who is normally assigned to the immediate work area mops it up. In our adjusted scenario let's switch the maintenance engineer normally assigned to an area to a field technician normally assigned to field sampling. For this scenario, we should not allow the field technician to clean up the spill, unless this worker has been adequately trained and is considered qualified. The recommended course of action for this field technician would be to leave the area and immediately contact the area supervisor and advise the supervisor of the situation.

Making this distinction is critical because there are different training requirements. Different exposure levels may apply depending on the phase of response. If post-emergency response is performed by an employer's own employees who were part of the initial emergency response, it is considered to be part of the initial response and not post-emergency response. Post-emergency response is defined under HAZWOPER as "that portion of an emergency response performed after the immediate threat of a release has been stabilized or eliminated and cleanup of the site has begun."

Let's look at another example. A 55-gallon drum containing flammable liquid was damaged during handling at a treatment, storage, and disposal (TSD) facility and is leaking. A worker calls the emergency response team, which arrives to manage the spill. While the team is performing its duties, a truck arrives with vacuum equipment to remove the spilled liquid. The team that managed the spill and the vac truck driver are considered to be part of the emergency response and should be trained accordingly [2].

HAZWOPER mandates a more conservative threshold for emergency response than an emergency defined under the DOE requirements. For example, a release of chlorine gas above the immediately dangerous to life or health (IDLH) level and moving through a building is an emergency situation under HAZWOPER. This is unlike an incidental release since the IDLH level has been exceeded. However, depending on the circumstances, the release may not be sufficient to require the declaration of an emergency under DOE [2].

The OSHA instruction on HAZWOPER generally refers to emergency responders as "employees who respond to emergencies." These emergency responders would include "employees from outside the immediate release area or other designated responders (e.g., mutual aid groups, local fire departments)" as well as "employees working in the immediate release area" to be designated as responders by the employer. This means that someone or some group of individuals from within the company having the emergency needs to be a liaison to the outside emergency responders.

For example, trained workers at a wastewater treatment facility are exchanging an empty 1-ton chlorine tank. A major leak occurs at the valve packing. The workers immediately evacuate the area and notify site authorities. In accordance with previously established procedures, the site emergency evacuation plan is activated, off-site emergency responses are summoned, and the incident command system is initiated. In this case, the workers who had been exchanging the tank are not required to be trained as emergency responders. But some in-house emergency team personnel would be expected to assist the emergency responders by identifying the release and providing other information and assistance to bring the situation under control.

The off-site emergency response forces require training and equipment in accordance with HAZWOPER and other applicable state or local criteria such as those promulgated by the National Fire Protection Association (NFPA).

11.2 APPLICABILITY OF SUPERFUND AMENDMENTS AND REAUTHORIZATION ACT

Title I of the Superfund Amendments and Reauthorization Act (SARA) regulations was issued to protect the health and safety of workers engaged in hazardous waste work. SARA Title I Section 126 (f) requires the Environmental Protection Agency (EPA) to issue standards for public employees in non-OSHA-approved plan States. The rules adopted by OSHA (29 CFR 1910.120 and 29 CFR 1926.65) and EPA (40 CFR 311) use the same basic concepts for worker protection of safety and health.

11.3 SARA TITLE III

SARA Title III, known as the Emergency Planning and Community Right-To-Know Act (EPCRA), was a law enacted to improve state and local government capacity to respond to emergencies caused by accidental releases of extremely hazardous substances. This law was designed to improve emergency preparedness and to give information to the public on hazardous chemicals made, used, or stored in their communities. It establishes requirements for industry regarding emergency planning and community right-to-know reporting on certain chemicals considered hazardous. This law builds on the EPA's Chemical Emergency Preparedness Program (CEEP).

SARA Title III is intended to help communities access information and be better prepared to deal with the presence of hazardous chemicals and releases of those chemicals into the environment. Through SARA, states and communities are encouraged to work together with facilities to improve hazardous materials safety and protect public health.

SARA has four major provisions or sections: emergency planning, emergency release notification, community right-to-know reporting requirements, and toxic chemical release inventory.

11.3.1 Emergency Planning (EPCRA Sections 301–303)

SARA requires the governor of each state to designate a state emergency response commission (SERC). SERCs include public agencies related to the environment, natural resources, emergency services, public health, occupational safety, and transportation. The SERC will designate local emergency districts and appoint a local emergency planning committee (LEPC).

The LEPC includes elected state and local officials. Besides elected officials, the LEPC could include police, fire, civil defense, public health, hospital, and transportation officials, as well as environmental experts and facility representatives. The LEPC requires the development of emergency response plans.

11.3.2 Emergency Release Notification (EPCRA Section 304)

Under this provision, facilities should notify the LEPC and consequently the SERC of any possible environmental release of specific chemicals. The specific chemicals referred to in SARA Title III are found on the Extremely Hazardous Substance List (40 CFR 355) and the Reportable Quantity List (the Comprehensive Environmental Response, Compensation, and Liability Act [CERCLA] Section 103 [a]).

Emergency notification should include chemical name and identification of the chemical by number; estimation of quantity released; time and duration of release; mode of release (air, water, or soil); known health risks associated with the emergency; applicable precautions; and name and phone number of a contact person. All emergency notifications require a written follow-up as soon as possible [2].

11.3.3 Community Right-To-Know Reporting Requirements (EPCRA Sections 311–312)

According to EPCRA, facilities should provide either an MSDS or a list of chemicals to the SERC, LEPC, and local fire department. If facilities choose to supply only a list, the list should include specific information including health hazards, fire hazards, reactivity hazards, and physical data for every chemical on the list. Although only a list is required, the additional requirements for specific information makes the submission of only a list a rare occurrence. The use of MSD sheets is one of the most commonly used tools to convey this information.

Facilities should complete an emergency and hazardous chemical inventory. Because inventories change, it is typical to see the inventory list contain ranges for the amounts of chemicals on hand. This inventory is to be submitted to the LEPC, SERC, and local fire department [2].

11.3.4 Toxic Chemical Release Inventory (EPCRA Section 313)

Under EPCRA, the EPA established an inventory of routine toxic chemicals that require emissions reporting. Facilities subject to Section 313 are required to submit a toxic chemical release inventory form or Form R for specified chemicals, which is completed on an annual basis and is submitted by July 1 of every year. Form R notifies public and governmental agencies about routine releases (releases that occur as a result of daily production use). Form R applies to facilities of ten or more employees in businesses (with standard industrial classification (SIC) codes 20 through 39) that manufacture or use certain toxic chemicals in excess of certain amounts.

The community HAZMAT emergency response plan can be a valuable source of information in developing site-specific emergency response plans and emergency action plans as required by HAZWOPER. This applies particularly to the need for coordination by DOE sites with off-site response personnel and agencies (e.g., mutual aid agreements and public alert mechanisms). EPA has provided guidance to communities and fire departments for identifying, acquiring, and maintaining HAZMAT response equipment and trained personnel appropriate for their locale.

The HAZWOPER standard requires employers to determine the potential for an emergency and develop response procedures accordingly. There are various methods for identifying and evaluating such hazards, each requiring different levels of resources and expertise [2].

11.4 EMERGENCY ACTION PLAN

An emergency action plan (EAP) is essentially an evacuation plan. It sometimes can be advantageous not to expect workers to respond to emergencies. When you consider the history of the number of deaths in confined spaces this point becomes clear. Prior to enactment of the confined space standard (29CFR 1910.146), according to statistics, more than two responders died for every entrant death. Typically, someone would see a coworker or friend in trouble within a confined space and go in after that individual without concern for personal safety. With better training and communication the statistics for deaths during responses to confined space emergencies have improved.

For confined space and other situations, personnel should be trained to walk away from the danger and not attempt to respond. The confined space standard is quite specific on how rescues are performed and how much preparation and training is required prior to confined space entry [3]. Personnel performing confined space work who have the proper training understand their roles in an emergency. Emergency planning for confined space entries is a requirement that most people take seriously because of documented serious consequences. But emergency training in general does not seem to strike as close to home for many workers.

So, training your personnel when to walk away becomes very important. And what they do when they walk away also becomes very important. Even the direction workers are expected to take can be important. The training performed and documentation of this training are also important.

Sites that intend to evacuate their employees from the danger area (and do not expect or allow any workers to participate in emergency response) when a release requiring emergency response occurs are required by OSHA to have an EAP as cited in 29 CFR 1910.38 (a):

- Emergency escape procedures and emergency escape route assignments. (A diagram or map works well.)
- Procedures to be followed by employees remaining to operate critical plant operations before they evacuate. (In operating processes this can be critical.)
- Procedures to account for all employees after emergency evacuation has been completed (head count in rally points).
- Rescue and medical duties for those who are to perform them. (These folks also require adequate training.)
- Preferred means of reporting fires and other emergencies. (If it is necessary to dial a code such as "9" before dialing 911, this should be clearly posted on every phone.)
- Names or regular job titles of persons or departments who can be contacted for additional information or explanation of duties of the plan (emergency call-out list).
- Pre-incident planning, coordination, and notification procedures with outside parties as required by 29 CFR 1910.120. (Start with your local fire department—it may have more expertise than you are aware of.)

11.5 EMERGENCY RESPONSE PLAN

An OSHA emergency response plan (ERP) is a written plan to prepare for and handle anticipated emergencies prior to the emergency. If employees are expected to respond to spills or releases requiring an emergency response, OSHA requires the development of an ERP that contains required elements as outlined in 29 CFR 1910.120 (q)(2) and (l)(3)(iv). The following are the minimum type of procedures:

- Pre-incident planning and coordination with outside concerns (for instance, local fire department or emergency response groups)
- Pre-emergency planning prior to operation (having meetings and tours with the responders to familiarize them with the site can be helpful)
- Personnel roles, lines of authority, training, and communication
- Emergency recognition, identification, and prevention
- Safe distances and places of refuge (rally points)
- Site security and control
- Evacuation routes and procedures
- Decontamination
- Emergency medical treatment and first aid
- Emergency alerting and response procedures
- Critique of response and follow-up
- PPE and emergency equipment
- Conduct of periodic drills

11.5.1 Emergency Response Organization

Development of procedures for handling emergency response, incident command protocols, and safety practices during an emergency is addressed in 29 CFR 1910.120 (q)(3). The following emergency response issues need to be addressed:

- Coordination and control of emergency responder communications
- Specific responsibilities with regard to use of engineering controls, hazardous substance handling procedures, and use of new technologies
- Self-contained breathing apparatus (SCBA) use requirements
- On-scene response, safety practice requirements, and safety official responsibilities
- Incident commander role, such as implementing decontamination procedures
- On-scene safety requirements for prebriefings for personnel, instructions for wearing PPE and for response duties, and health and safety precautions for support personnel

An incident command system (ICS) is an organized approach to effectively control and manage operations at an incident involving hazardous substances, regardless of size. Implementation of the ICS is required by the HAZWOPER standard. An effective ICS should avoid confusion, improve safety, organize and coordinate actions, and facilitate effective management at the scene of an incident. The basic elements of an ICS include the following:

- Consolidated action plans
- Modular organization
- Incident commander
- Unified command structure
- Manageable span of control
- Integrated communications
- Predesignated facilities
- Comprehensive resources management

The individual in charge (the incident commander) of the ICS is the senior HAZMAT official responding to the incident. The incident commander has full authority to carry out his or her responsibilities and priorities, which include protection of personnel, property, and the environment at the emergency scene. An ICS must ensure that an incident commander is appointed and a system is established to address the practical aspects of on-scene response, responder safety, and return to normal operations.

When off-site emergency response groups are expected to provide primary support or any backup support for a hazardous material emergency, advance coordination with those groups regarding the ICS is needed. Site and off-site emergency response plans and procedures for on-scene incident response and command should be coordinated to make certain that it is understood who will be the individual in charge of on-scene incident response. Sites with trained and equipped responders will typically provide the on-scene incident commander with mutual aid responders reporting to this individual [2].

11.6 EMERGENCY EQUIPMENT AND PERSONAL PROTECTIVE EQUIPMENT

Areas of guidance in the HAZWOPER standard not specified in DOE orders include but are not limited to the following:

- **SCBA use in emergency response:** 29 CFR 1910.120 (q)(3)(iv) requires that a positive-pressure SCBA be used "while engaged in emergency response, until such time that the individual in charge of the ICS determines through the use of air monitoring that a decreased level of respiratory protection will not result in hazardous exposure to employees." If the incident commander believes that hazards are not adequately characterized, he or she must order the use of positive-pressure SCBAs.
- **Approved cylinders:** 29 CFR 1910.120 (q)(3)(x) requires that "approved SCBAs may be used with approved cylinders from other approved SCBAs provided such cylinders are of the same capacity and pressure rating." Interchangeable cylinders become important during emergencies. For respiratory protection during nonemergency situations the NIOSH approvals require that certain components are not "mixed and matched" or the approval may not be allowed.
- **Chemical PPE:** In a fire or thermal energy hazard, PPE worn by responders should meet, at a minimum, the criteria in 29 CFR 1910.156 (e), "Fire Brigade Standard," requiring turnout gear. In conditions where skin absorption of a hazardous substance may result in substantial possibility of immediate death, serious illness, or injury or impaired ability to escape, totally encapsulated chemical protective suits should be used. It is vital to keep heat resistance of the totally encapsulated suits and the heat resistance of any PPE used underneath or in conjunction with the totally encapsulated suits in mind any time there is a thermal hazard.

Information gathered at the site characterization stage of an emergency response operation influences all other aspects of the response (e.g., delineation of contamination zones). Based on characterization of the emergency site, the incident commander is responsible for implementing appropriate emergency response operations and making certain that appropriate PPE is used, recognizing that turnout gear may not be appropriate for chemical exposure emergencies.

The incident commander may rely on visual observation of placards, labels, and manifests and information gathered during the response. Obtaining air measurements with monitoring equipment for toxic concentrations of vapors, particulates, explosive potential, and the possibility of radiation exposure is important for determining the nature, degree, and extent of the hazards [2].

11.7 MEDICAL SURVEILLANCE

OSHA's HAZWOPER standard contains specific requirements with regard to medical surveillance of emergency response team members and provision by the physician of a written medical report to the individual. As cited in the OSHA instruction, if response activities involve infectious materials, the site is to comply with 29 CFR 1910.120 (q) and may also have to comply with 29 CFR 1910.1030, "Bloodborne Pathogens." If there is a conflict or overlap, the provision that is more protective of employee health and safety applies.

Members of a HAZMAT team are to receive baseline physical examinations to certify their physical ability to perform assigned duties, including the ability to work in the particular PPE that may become necessary. They should be provided with medical surveillance annually and after a hazardous substance exposure. HAZMAT emergency responders must participate in an ongoing medical surveillance program. The employer is required to furnish the employee with a copy of the physician's written opinion indicating medical results and whether the employee is capable of wearing the PPE that may be required while working with hazardous substances.

Any emergency response employee who exhibits signs or symptoms that may have resulted from exposure to hazardous substances during an emergency incident is to receive medical consultation. The responder's employer is to provide to the physician a description of the employee's duties as they relate to the individual's exposure, the responder's exposure level, a description of any PPE used, and information from previous medical examinations of the employee that is not readily available to the examining physician. The responder is to be furnished a copy of a written opinion from the attending physician, including the physician's opinion on any detected medical conditions that would place the employee at increased risk. This written opinion must include the physician's recommended limitations on the employee's assigned work, and the results of the medical examination and tests [2].

11.8 EMERGENCY MEDICAL TREATMENT, TRANSPORT, AND FIRST AID

The emergency response organization at your facility should develop and maintain an information and communication system with local medical centers for treatment beyond site capability for injured, contaminated, or irradiated personnel. Coordination with hospitals or other medical care providers prior to emergencies is very important [2].

REFERENCES

1. *Occupational Safety and Health Guidance Manual for Hazardous Waste Site Activities.* Prepared by National Institute for Occupational Safety and Health (NIOSH), Occupational Safety and Health Administration (OSHA), U.S. Coast Guard (USCG), U.S. Environmental Protection Agency (EPA), October 1985, p. 12-1.
2. *Handbook for Occupational Health and Safety During Hazardous Waste Activities.* Office of Environmental, Safety and Health Office of Environmental Management, 1996, pp. 10-1–10-9.
3. 29 CFR 1910.146 OSHA "Confined Space Standard."

Appendix A

OSHA Site Audits

The authors believe that although the following information is specific to superfund sites, the general findings are universal. As you review this information, you will notice some areas of bold print. Within these bolded areas, the authors have added their own analysis, comments, and lessons learned while performing field activities at sites of all sizes that deal with hazardous materials.

Except for the bolded sections, the following information was taken directly from a report entitled "EPA/LABOR Superfund Health and Safety Taskforce: OSHA Audits of Superfund Sites 1993 to 1996," dated August 25, 1997. The report is lengthy, so the authors have only included parts of the report they felt to be pertinent to the subjects within the main body of the book.

I. INTRODUCTION

Beginning in the early 1990s, the Occupational Safety and Health Administration (OSHA) and the U.S. Environmental Protection Agency's (EPA's) Office of Emergency and Remedial Response (OERR) have participated jointly, through an interagency agreement, in Superfund site health and safety audits. These audits are an attempt to ensure effective occupational safety and health oversight of Superfund remediation operations. As part of this initiative, OSHA conducts in-depth safety and health evaluations of selected Superfund sites using various remediation technologies, including incineration, in-situ vitrification, soil washing and lead leaching, and low-temperature enhanced volatilization. These evaluations or audits are not enforcement actions. Rather, they are intended to assist the site contractors and Federal and State oversight personnel to improve their understanding and implementation of OSHA requirements and recommendations for site health and safety. Whenever an audit is conducted, occupational safety and health oversight personnel from the Federal and/or State office are invited to participate.

The information that follows summarizes the findings of audits and site safety and health plan (SSAHP) reviews performed for eleven

Superfund sites between 1993 and 1996. Although a major objective of these evaluations is to assess compliance with OSHA's Hazardous Waste Operations and Emergency Response (HAZWOPER) standard (29 CFR 1910.120), they also seek to evaluate the overall adequacy of each facility's safety and health program, as implemented by the contractors operating at each site, and to identify any factors that contribute to reduced program effectiveness. Emphasis falls on evaluating each employer's safety and health standard operating procedures (SOPs) and the adequacy of task- and of specific hazard analyses and emergency response programs. In addition, the evaluations extend to such areas as heat stress mitigation strategies, confined-space programs, and process safety management approaches used during remediation operations.

OSHA conducted site inspections and SSAHP reviews for eight of the eleven sites discussed in this summary report. Inspections of these sites were conducted by teams of four to six OSHA personnel. The audit teams generally spent three or four days at the site interviewing employees, safety and health personnel, union representatives, and site management personnel to evaluate the effectiveness of safety and health program implementation; conducting walk-through inspections to observe and document site conditions, operations, and safety and health program deficiencies; collecting wipe samples of work surfaces and in some cases, wipe samples of employees' skin; and reviewing each site's written safety and health plan, including the emergency response plan, operation-specific hazard analyses, and other relevant written safety and health programs and records.

The other three sites discussed in this report did not undergo site audits. OSHA did, however, thoroughly review their written SSAHPs and related documents.

III. DESCRIPTION OF AUDITED SUPERFUND SITES

When EPA is unable to identify the responsible party for a Superfund site, or cannot reach an agreement with the responsible party, EPA performs the necessary remedial design and remedial action (RD/RA) activities. In such cases, EPA chooses between two contracting mechanisms to conduct the RD/RA: EPA may provide direct oversight of the RD/RA contractor under the Alternative Remedial Contracting Strategy; or EPA may request that a RD/RA be administered and implemented by the U.S. Army Corps of Engineers (USACE) or Bureau of Reclamation (BUREC) under an Interagency Agreement with EPA. In some instances, a state agency will assume responsibility for RD/RA and use its own contracting mechanisms. Under any of these circumstances, the agency that issues the competitive bid contract provides oversight of the prime contractor selected to perform cleanup activities. The prime contractor is responsible for implementing cleanup procedures in accordance with the terms of the contract and for developing and implementing a safety and health program for the site. The prime contractor may procure the services of a number of subcontractors that specialize in various aspects of the cleanup activity such as operation of an incinerator or dredging.

Although we will be providing the site and site location, we will not be identifying prime contractors or sub-contractors.

American Thermostat

Remediation activities at American Thermostat included the excavation and thermal treatment of over 13,000 cubic yards of soil and sediments contaminated primarily with perchloroethylene, trichloroethylene, and solvents. The soil was excavated and treated using a thermal treatment unit called the low-temperature enhanced volatilization facility (LTEVF). The performance test for the site's thermal unit had just been completed at the time of the inspection, so there was limited activity.

Arlington Blending and Packaging (Arlington)

Arlington is the 2.5-acre site of a company that formulated technical grade chemicals, primarily pesticides. The site contained concrete pads from previously demolished buildings, non-native gravel, and a concrete block/sheet metal building. Site investigations identified chlordane, endrin, heptachlor, heptachlor epoxide, pentachlorophenol, and arsenic at concentrations above background levels in site soils and groundwater. Remediation activities included the excavation of soil and concrete slabs, demolition of the remaining building, pretreatment and stockpiling of soil, low-temperature thermal desorption of soil contaminants, handling of treated soil, site stabilization (if necessary), and wastewater treatment.

Baird & McGuire (Baird)

Baird is the 20-acre site of a former chemical mixing and batching company. Poor waste disposal practices resulted in the contamination of groundwater, soil, the municipal water supply, and a brook adjacent to the site. Over one hundred contaminants, including chlorinated and nonchlorinated volatile organics, heavy metals, pesticides, herbicides, and dioxins, had been identified in site soil and groundwater. Remediation activities included soil excavation and incineration, and groundwater treatment (the audit focused on the soil excavation and incineration portions of the project). Specific activities included site preparation, construction of a facility to house incinerator operations, incinerator installation, excavation and incineration of 155,000 cubic yards of soil, backfilling, and land restoration. The incinerator at this site is a portable unit that can be disassembled at the conclusion of site operations and reassembled at another site.

Brio Refining Site (Brio)

The Brio refining site is approximately 58 acres in size and is the location of a former chemical production, recovery, refinery, and regeneration facility. The site includes closed impoundments into which hazardous substances were disposed in bulk, storage tanks, and approximately 1,750 drums of hazardous substances. Remediation activities included the excavation and incineration of contaminated soil, installation of protective liners around selected pits, and the installation of a groundwater extraction system adjacent to a gully.

Manistique Harbor

The Manistique Harbor site is a dredging project located on the Manistique River north of Manistique Harbor on Michigan's upper peninsula. A local paper manufacturing company used the river as a source of cooling water and as a discharge point for wastewater and other mill effluent. The paper manufacturing company has been identified as the source of polychlorinated biphenyls (PCBs), the contaminant of concern, in the river sediment. Portions of the river are currently being dredged, and the contaminated sediment is being shipped to a remote disposal site. The waste sediment consists of wood chips, dirt, and sand. The wood chips contain the majority of the PCB contamination, which allowed for efficient waste segregation. Water removed with the sediment was treated at an onsite wastewater treatment plant constructed for the project.

Metaltec/Aerosystems (Metaltec)

Metaltec is the 16-acre site of a small metal casing fabrication plant and includes an unlined lagoon used for dumping waste solvents from the plant's operations. The waste solvents contaminated both the soil and groundwater on site and were the focus of the remediation efforts. Four parcels of land on the site were originally identified for remedial action. Soil remediation was completed on three of the four parcels in prior years; the remaining parcel included the area encompassing the unlined lagoon.

Remediation activities included site mobilization (i.e., installation of trailers, utilities, and equipment; clearing and grubbing; grading roads; and construction of decontamination facilities, drainage pump stations, and a water treatment system), soil excavation, thermal processing of 7,700 cubic yards of soil, backfilling and regrading the excavated area, and site demobilization.

North Cavalcade

Remediation activities at North Cavalcade included the installation, operation, and closure of a bioremediation system to treat contaminated soil.

Sand Creek Corridor Site (Sand Creek)

Sand Creek is located in an industrial area that contains petroleum and chemical production and distribution facilities, trucking firms, warehouses, and residences. The site contains contaminated soil, sediment, groundwater, surface water, and buildings. The site audit addressed operations at three of the site's operable units which contain contaminated groundwater and soil contaminated with volatile organics, pesticides, herbicides, and metals. Remediation activities included vacuum extraction of volatile organics, excavation and containerization of soils, dismantling and demolition of buildings and structures, and drilling in support of groundwater monitoring efforts.

Twin City Army Ammunition Plant (Twin City)

Twin City is the 10-acre site of a former U.S. Army ammunition production facility. Attempted destruction of off-spec or damaged ammunition contaminated the soil with lead and other heavy metals including antimony, cadmium, chromium, copper, mercury, and nickel. Soil decontamination involved a new soil-washing and lead-leaching technique designed to generate no waste streams.

Vertac

Vertac was divided into two parcels: Parcel 1 contained abandoned herbicide production facilities and equipment and current groundwater treatment facilities; Parcel 2 contained an incinerator and staging areas for drummed waste. The incineration operation involved thermal destruction of about 28,000 drums of waste 2,4-D and 2,4,5-T and 2,4,5-T still bottoms. Known contaminants included toluene, chlorobenzenes, chlorinated phenols, acids, dioxin, and the pesticides 2,4-D, and 2,4,5-T.

Wasatch Chemical (Wasatch)

Wasatch is an 18-acre site that formerly hosted operations for warehousing, producing, and packaging industrial chemical products, including pesticides, herbicides, fertilizers, and other industrial chemicals and cleaners. The site includes accumulated debris, as well as sludge and soil that are contaminated with semivolatile organics, volatile organics, metals, and various pesticides and herbicides. Remediation activities included the consolidation of contaminated site debris, sludge, soil, and dioxin wastes into the former evaporation pond located on the site; destruction of the organic chemicals in these materials using in-situ vitrification; excavation and land framing of toluene- and xylene-contaminated soils; installation of a groundwater extraction and treatment system; and, as necessary, construction of a groundwater containment system and treatment facility.

At the time of the audit, remedial action was nearly complete.

IV. AUDIT RESULTS

OSHA found multiple deficiencies in the design, management, and implementation of safety and health plans at all the sites that it visited and in all of the plans reviewed. These deficiencies fell into twelve functional areas and were often common to all the sites. A discussion of the findings specific to each functional area follows. Note, for the remainder of this report, that the identities of the sites are masked and are referred to by a randomly assigned letter designation (Site A–K).

A. Safety and health supervisors at the site must be given the authority to exercise their judgment in matters of employee safety and health. Management decisions related to safety and health must reflect the judgment of such individuals.

Perhaps the most essential component of the safety and health program at a hazardous waste site is the development, management, and implementation of the program by a competent site safety and health supervisor who has the authority to make timely decisions as worksite conditions rapidly change. The safety and health supervisor must also have the flexibility to conduct any investigations necessary to fully characterize the hazards to which employees may be exposed and to ensure that the safety and health program is effective in mitigating those hazards. The need to grant appropriate authority to the site safety and health supervisor is addressed in paragraph (b)(2)(I)(B) of 29 CFR 1910.120 (HAZWOPER).

Safety plans must make clear site-specific responsibilities and lines of authority. Examples should be used in the plan so that it clearly spells out what must take place in a variety of situations. An organizational chart should become part of any larger site's plan. The dotted line and solid line responsibilities give on-site personnel a clear indication of what is expected of them, and to whom they should report unsafe or unhealthy situations.

Sites B and H had on site qualified safety and health supervisors with the authority to exercise their judgment in matters of employee safety and health. At Site H, however, the related contractual agreement between the prime contractor and the lead government agency limited the health and safety manager's authority in areas such as downgrading PPE levels by establishing inflexible minimum PPE requirements. As a result, the PPE requirements used on the site at the time of the audit seemed excessive in light of site monitoring data and hazard determinations. This use of excessive PPE unnecessarily increased the risk of heat stress and other PPE-related hazards.

These types of situations remain problematic for a variety of reasons. First and foremost, safety professionals write plans for and are concerned with avoiding accidents. When PPE is being planned, the author of the plan will typically take a conservative approach toward PPE. Most times, the plan's author has not seen the site and does not have in-depth information or pictures of the actual working site.

Budgets are also a concern. Safety plan authoring, like any other activity, usually has a fixed budget. Many times a reduced budget in authoring the plan results in failure to adequately plan. In an effort to protect workers and still remain in the budget, "catch-all" PPE is used. Catch-all PPE typically includes saranex, booties, taped seams, glove liners, viton outer gloves, and a face shield. Although the plan may come in under budget, the bill for the PPE is usually excessive.

In addition to budgetary constraints, safety people have a tendency to want to place workers "in a bubble." If the authors are just a bit unsure of what will be encountered, they tend to overcompensate with PPE. These types of plans typically do not include downgrade statements, but they will allow workers to upgrade without approval. In order to downgrade, the plan usually requires workers to get numerous persons' approvals in writing or go through a formal amendment process. Although amendments to plans have been made, folks are reluctant to do so for a variety of reasons. The bottom line is, no amendments are made, the PPE levels remain the same, and workers are forced to work in levels of protection that are not warranted.

Working in more protective levels of protection than conditions truly warrant, as noted in the above audit finding, is commonplace and still happens today. There is an old saying, "Better safe than sorry"; however, when all items are considered, too much protection in the form of PPE can create a dangerous situation.

The safety and health programs at Sites F, G, J, and K did not establish clear lines of authority or communication in the area of site safety and health. At Site K, neither contractor's SSAHP established a chain-of-command for the site. Although the plan listed the safety and health responsibilities of the project manager, industrial hygienist, and site safety officer (SSO), it did not describe the lines of authority and communication among them. At Site G, the relationship between the health and safety officer (HSO) and SSO was not clear, and it appeared that no single individual had overall authority for site safety and health. Site J's site management and health and safety management were handled from the headquarters office in another city. Site J had an operations manager on site whose responsibilities included day-to-day operations and compliance with the SSAHP, but this individual did not have a background in safety and health. Three or four years prior to the audit, a site health and safety manager was employed at the site, but his responsibilities were not reassigned after his departure and employee interviews indicated that since his departure, it was unclear who was overseeing worker safety and health. As a result, Site J workers developed a "make do" attitude regarding safety and health oversight. For example, Site J workers signed one another's confined space permits. At Site F, the SSAHP referred to safety and health personnel who were not mentioned in the organizational structure, making lines of authority ambiguous.

Clarity in planning makes successful execution much easier than does ambiguity. An organizational chart should be included any time there is the potential for ambiguity. Everyone on site must know who is ultimately in charge of safety at any one time. This does not mean that personnel cannot wear more than one hat at one time, but does mean that everyone on site must be in the loop as far as who is in charge at which time.

At Sites E and I, each prime contractor had designated a safety and health supervisor, but these individuals did not have the knowledge and/or authority necessary to implement the site safety and health plans or to verify compliance. The prime contractor's SSAHP at Site I designated a HSO, who was responsible for implementing the SSAHP and had the authority to shut down operations that pose a potential threat to site personnel. This HSO's area of expertise, however, was construction management. Accordingly, the audit team was concerned that he did not possess the necessary qualifications to manage the site safety and health program and felt that his lack of training could result in an inability to detect important safety and health deficiencies. Similarly, the site management personnel of the Site I subcontractor seemed unfamiliar with equipment maintenance, operating procedures, and audit procedures. As a result, most of the requirements related to these procedures had not been implemented.

Having qualified persons in the role of health and safety officer is required. How to determine minimum qualifications at each site is a site-specific task. It would depend on site activities, required and anticipated levels of protection, training requirements, general job knowledge, and a variety of other factors. Sometimes choosing a qualified SSO can be quite difficult. The authors are in agreement that personally contacting references is very important.

However, in the recent past we have found that references must be considered very carefully. When interviewing project managers regarding their SSOs, some would give an excellent reference to an SSO, saying that the SSO "did not give them any problems" or "got along well with everyone." Both of those attributes are excellent; however, these excellent attributes could indicate that the SSO really did not perform the job well. If they did not give anyone any problems, it is likely that the site they were working on was perfect, or they were not doing as good a job as one might believe. In either case, having set criteria of minimum qualifications should be either in the job specifications or in the safety plan.

The Site I subcontractor also lacked a backup safety and health supervisor fully trained in site safety and health management. The acting health and safety specialist (HSS) at Site E was a site worker who had held the position for one week. This individual stated to the audit team that he did not meet the corporate or SSAHP qualifications required to hold that position. Apparently, while the Site E contractor was waiting for the results of the LTEV performance test and the unit was not in operation, site management determined that less health and safety oversight was needed.

Site conditions and work activities change constantly. Change, and what effects change have on health and safety, should be included in every health and safety plan. If not, the plan needs to be amended to reflect current site conditions. Individuals' names for key on-site positions should be listed in each safety plan. If the personnel change, the plan should be amended. But prior to the amendment, a review of the replacement persons' training and qualifications should take place to ensure that qualified persons are chosen.

The regular HSS was on a month's leave. Responsibility for management of site safety and health at Site A was assigned to several individuals including the response manager, the Technical Assistance Team (TAT), health and safety officer (HSO), and the on-scene coordinator. It was not clear at the time of the audit which of these individuals actually fulfilled the role of full-time HSO, nor was it clear how these individuals would coordinate and communicate on overlapping health and safety issues.

In addition, several site HSO responsibilities identified in the health and safety plan were not completed by any of the individuals listed in the organizational section.

The site-specific safety and health plan (SSAHP) must include procedures for implementing and enforcing safety and health rules for all persons on site, including employers, employees, outside contractors, and visitors.

To maintain adequate site control, the site safety and health supervisor must have the authority to enforce the SSAHP's rules on any individual present at the site, whether that individual is an employee or an outside contractor. If there is more than one SSAHP (i.e., if each contractor develops its own), OSHA considers it essential that the plans be integrated and enforced consistently to ensure that on-site personnel have a clear understanding of safety and health expectations, lines of authority, and emergency response actions.

The audit team found that at Sites B and H, safety and health personnel had sufficient authority in most cases to thoroughly implement the safety and health plans. To facilitate safety and health compliance, anyone who entered Site H, including delivery personnel, was required to view a brief video that summarized the site history and remediation operations, identified the hazardous substances monitored on site, and described site evacuation procedures. Site B's SSAHP did not mention specific subcontractors and their roles and responsibilities; however, it did indicate that the prime contractor had oversight responsibility for all safety and health activities and the authority to discontinue or modify site operations when unsafe conditions were detected.

Videos have been shown to be effective in many situations. However, having a person in authority who gives orientation has certain advantages. These advantages are similar to those that a student who has a live professor enjoys versus the student who takes a correspondence course. If the student feels the need to ask a question or needs a point clarified, this can be handily accomplished with a live professor. This is usually not as easily done in a correspondence course. As far as showing delivery persons a video, we feel it is more advantageous not to orient certain persons, but to restrict activities of delivery persons and the like to support or clean zones. Using this approach, you spend a lot less time in orientation and keep unnecessary personnel out of zones where the potential for exposure exists.

The SSAHPs at Sites G and K did not establish clear lines of authority between the contractor and subcontractors. At Site K, neither the contractor's nor the subcontractor's SSAHP established a chain-of-command or lines of communication for the site, and neither plan mentioned the other contractor, despite the substantial impact each had on the other's operations. Additionally, management personnel and employees at Site K did not have a thorough understanding of their respective roles and responsibilities during site emergencies. At Site G, it was not clear whether subcontractors would be performing any portion of the work and, if so, how their respective SSAHPs would be integrated. The SSAHPs at Sites F and I lacked organizational information, such as clear lines of authority and communication, necessary to ensure the implementation and enforcement of safety and health rules for all persons on site. The prime contractor's SSAHP at Site I contained a corporate hazard communication policy but not a site-specific one. The Site I subcontractor's SSAHP had not been revised since 1993 and did not reflect current site organization or lines of authority, nor did it contain site-specific detail about personnel roles and responsibilities or procedures for how site contractors would be informed of hazards. At Site F the organizational structure, including lines of authority and communication, was not clear. The SSAHP repeatedly referred to personnel who were not listed in the organizational structure.

The use of personnel not mentioned in the plan appears to be commonplace. Field personnel can change on some projects. The plan should contain some guidance on qualifications on personnel and should be used as a training aid for the "replacement" personnel. It would be advisable to immediately reissue the plan to include the current personnel roster, but as OSHA has discovered, this is sometimes not done as quickly as one might hope. The authors believe that the situation mentioned above can likely be found on many sites even today.

For example, the organizational structure did not contain information about "field crew members" who elsewhere in the plan were assigned specific safety and health or emergency response duties.

Site C's SSAHP clearly stated that safety and health requirements described in the plan apply only to the employees of the prime contractor and subcontractor, and to visitors under the direct control of the contractor. As a result, the SSAHP did not cover other individuals on site such as EPA personnel; state and local government personnel; or employees, representatives, or contractors of the Potentially Responsible Party (PRP).

A visitor policy should be established in the SSAHP. Visitors and representatives from government and other organizations should be specifically mentioned. Zones that visitors can tour, and the circumstances under which the visitor may tour a zone, should be specified in the plan. If someone needs to go into an exclusion zone, those persons must have the appropriate orientation, safety training, medical clearance, and meet any other requirements mentioned in the SSAHP.

Typically, visitors have little or no reason to enter exclusion zones. An observation deck or other area out of the exclusion zone can usually allow any visitor (with or without the aid of binoculars) to observe work activities as necessary. Video cameras have been used successfully to show

visitors or representatives any activities within the exclusion zone for which there is an interest.

C. The safety and health program must effectively ensure that ongoing task-specific hazard analyses are conducted so that the selection of appropriate PPE can be made and modified as conditions warrant.

The OSHA standard (29 CFR 1910.120) mandates that site safety and health programs require task- and operation-specific hazard analyses be conducted at the site. These analyses are intended to ensure a comprehensive and systematic approach to hazard anticipation, recognition, and evaluation at hazardous waste sites. Since work operations and site conditions change at different stages of the remediation process, the potential hazards associated with each operation must be reevaluated periodically to ensure that employees receive appropriate protection.

For example, as work progresses, all information and data on employee exposures obtained to date should be incorporated into the analysis to enhance and refine the evaluation. The results of air monitoring are an important source of site-specific information used for hazard analysis. The requirement to conduct task- and operation-specific hazard analyses and to incorporate the results of such analyses into the site-specific SSAHPs is contained in paragraph (b)(4)(ii)(A) of the HAZWOPER standard. Paragraph (h) of the standard also requires that employees be monitored to ensure adequate characterization of their exposures and that the results of all exposure monitoring be fed back into the hazard analysis process to ensure continuing improvement in site planning and procedures.

The OSHA audit teams found program deficiencies in two related areas: the development of task- and operation-specific hazard analyses; and the conduct of monitoring programs designed to characterize employee exposures to hazardous materials. These deficiencies are discussed in more detail below.

Task- and Operation-Specific Hazard Analyses

Six of the eleven sites evaluated (Sites A, B, F, G, I, and K) identified generic remediation hazards in their SSAHPs but did not address the hazards associated with site- and operation-specific tasks. For example, the SSAHP for Site B broadly described hazards such as "the potential for inhalation, ingestion, contact, and absorption of contaminants" or "heavy equipment and general construction hazards." The plan did not describe specific hazards (i.e., levels of specific chemical contaminants, the hazards related to the use of specific types of equipment) associated with specific workplace activities and their related control measures. These general analyses do not provide employees with sufficient information to enable them to work safely, nor do they enable the employer to determine the types and levels of controls necessary to protect workers from health and safety hazards.

In the absence of site-specific exposure information, the selection of PPE at these sites did not appear to be based on the performance characteristics of the PPE relative to the hazards and potential hazards. For example, at Site G, no rationale was provided for using modified Level C PPE for a few jobs for which use of a half-mask respirator is permitted. Additionally, the Site F site plan did not contain any site-specific PPE information that employees could use for site tasks and operations.

The lack of specificity and the potential reasons for a lack of specificity were discussed in the previous section. The SSHAP should be specific enough to discuss site-specific health and safety hazards. A rationale for the use of PPE should be provided, especially when upgrading levels of protection.

Employee Exposure Monitoring

Site C had one of the most comprehensive employee exposure monitoring programs of the sites reviewed. The Site C SSAHP described an extensive exposure assessment program that included both real-time monitoring with direct-reading instruments and personal air sampling. It also established action levels for explosive atmospheres, organic vapor concentrations, dust concentrations, and noise levels that trigger PPE use or evacuation. The SSAHP specified that personnel sampling is to be performed daily on all contractor personnel, unless monitoring data indicated that a lower frequency (e.g., once per week) would be acceptable.

Personal sampling was conducted for the eight contaminants, which appeared to be the most common on site. Site characterization data, however, indicated the presence of other contaminants for which OSHA has monitoring requirements and/or PELs. The SSAHP mentioned the use of a data management system for recording monitoring results but did not describe procedures for data analysis or the use of monitoring data to revise the sampling plan.

The contractors at Sites H and K also conducted personnel monitoring activities; however, neither used this data to determine appropriate PPE levels. At Site K, real-time monitoring for total organics was conducted daily by both contractors; however, the air sampling records and corresponding results were not stored together on site, making it difficult to correlate results with sampling information, and therefore, to accurately assess risks to employees. PPE determinations at Site K were not always based on monitoring results. For example, both contractors used a modified Level C with full-face respirator although real-time air monitoring for volatile organics seldom showed detectable airborne levels of contaminants.

In addition, the use of chemical protective clothing was not supported by air or surface contamination monitoring to determine the potential for dermal exposure and the appropriate PPE.

The contractor at Site H had established area and personnel sampling consistent with HAZWOPER requirements. A photo ionization detector (PID) and a real-time aerosol monitor (RAM) were used on a daily basis to screen for potentially hazardous levels of contaminants. On a weekly basis, personal air samples were collected and submitted for laboratory analysis. PPE requirements, however, were often not based on this data because the oversight agency had established inflexible minimum PPE requirements. The audit team found many of the PPE requirements on Site H to be excessive in light of site monitoring data and hazard determinations.

A conservative initial approach makes sense when working with hazardous materials. This is especially true when you are dealing with a substance that you are not completely sure of such as those on hazardous waste sites. Hazardous waste sites can become extremely challenging when substances are discovered that were not previously known. This discovery of unknown substances as hazards should not take place when dealing with hazards in a manufacturing facility or more controlled environment. This dealing with the unknown, or digging up something totally unexpected, can make hazardous waste work extremely challenging.

However, after becoming familiar with site hazards as best as one can, along with analytical data, the level of protection should be reexamined. This reexamination should be built into the plan, and specific criteria for downgrading levels of protection or the type of protective equipment should be considered. Even a downgrade in the type of coveralls required can make a large difference in worker heat stress load. The ability of tyvek to "breathe" (as opposed to saranex) is very desirable when considering worker comfort and heat load. These options, and when they become viable, should be anticipated beforehand and placed in the SSAHP.

Contractors at Sites B, D, G, I, and J had incomplete sampling practices and as a result were not able to evaluate PPE levels based on monitoring data. For example, both contractors' SSAHPs at Site I lacked provisions for monitoring site hazards such as metals, pesticides, herbicides, and semi-volatile organic compounds (SVOCs) that could not be evaluated with a PID. Since worker exposures to the range of hazards on site had not been characterized, PPE was not selected based on its performance relative to the nature and level of site hazards.

The sampling and monitoring frequencies specified in the Site B site contractor's SSAHP were not consistent with HAZWOPER requirements since the plan based monitoring frequency on prior sampling results alone and did not consider other factors such as the performance of new or different tasks. The SSAHP did not include wipe sampling among the specified monitoring procedures despite the fact that most of the site's identified contaminants pose significant dermal exposure hazards. In addition, the plan included a monitoring technique that appeared to be unsuited to evaluation of airborne contaminant levels at the site. The Site B SSAHP did address ambient air monitoring and personnel sampling; however, those requirements did not cover potential exposures to each of the hazardous substances identified on the site. As a result, site characterization data provided in the SSAHP did not provide the site-specific airborne exposure levels necessary to determine an appropriate level of PPE for each specific task.

The reason behind the incomplete sampling results needs to be determined. Once the determination has been made and the corrective actions implemented, the necessary results should be forthcoming. Hopefully, if follow-up or follow-through has been performed, these situations are corrected before they become critical. Unfortunately, a plan is only as good as the folks who execute the plan. Writing a plan is a necessary first step in ensuring safe on-site work activity. However, if an underdeveloped safety culture exists on site, difficulties will quickly be realized. However, building a safety culture is a whole other subject that is not covered in this book.

The SSAHP for the contractor at Site G did not specifically identify the chemical contaminants for which employee exposure monitoring would be conducted, with the exception of some indicator tube samples. An appendix of the SSAHP contained NIOSH sampling methods; however, with the exception of measuring metal fume exposure during hot work, it described no program for personal sampling of employee exposures. It appeared that all employee exposure sampling was performed with direct-reading instruments. Site G's program did not address how the results of monitoring would be reported to affected employees, nor did it establish exposure limits or triggers for PPE selection and use.

An acceptable SSAHP will specifically address hazards of concerns, action levels, and protective measures and techniques workers must use. This will include the use of direct reading instruments coupled with TWA-type sampling when warranted.

Monitoring records maintained at Site J provided no indication that any contractors were fulfilling regulatory or contractual requirements for monitoring. In an interview, an employee explained that the lack of personal monitoring was based on ambient air sampling results indicating that the levels of contaminants were too low to warrant personal sampling. The monitoring results available on site at the time of the audit were inadequate to justify this conclusion. At Site D, PPE selection was not based on monitoring data. For example, PPE was required in the support zone, although no surface sampling had been conducted to determine the need for PPE in that area.

The PPE chosen to protect workers should be justified. Part of the justification for wearing PPE is sampling results. The SSHAP should include the specifics as to how and when this will be accomplished.

Exposure monitoring had not been conducted at Site A. The decision not to conduct monitoring was appropriately documented in the site safety plan and was based on the nature of the contaminated material. The sediments were always saturated, and the likelihood of airborne concentrations of PCBs was extremely low. Area samples had been collected at the waste pad where the material was air-dried prior to shipping. Sample results were below the detectable limits.

The site safety plan indicated, however, that noise monitoring was required, but neither area noise surveys nor personnel dosimetry had been conducted.

Not performing monitoring that is required by the SSHAP may have potentially serious consequences. For noise monitoring, guidance should be written into the plan that specifies when noise monitoring will be performed. For instance, how should one handle the following situation? Let's say you are on site but you do not have a sound level meter. At what point do you need to monitor? One rule of thumb for this situation is as follows: if you cannot conduct a normal conversation with fellow employees that are within three feet, you are likely at or above 85 dBA. Your plan should state this. And it should state when and if the site would need to have a sound level meter on site.

D. Task-specific hazard analyses must lead to the development of written standard operating procedures (SOPs) that specify the controls necessary to safely perform each task. Detailed hazard analyses conducted for each site task and operation provide the basis for developing SOPs to protect employees from safety and health hazards. Written SOPs provide a mechanism for informing employees of procedures that ensure their safety and for enabling management to enforce hazard control procedures.

Requirements for written safety and health SOPs are found in paragraphs (b)(4)(I) of the HAZWOPER standard. None of the sites audited had developed comprehensive, site-specific SOPs that specify the controls necessary to complete each task. The contractor's SSAHP at Site B did discuss generic task-related hazards and SOPs; however, the task-specific SOPs lacked detail. For example, the SOPs did not specifically identify the site locations where hazards were likely to occur, nor did they specify the appropriate types of controls.

The prime contractor's SSAHPs for Sites D, F, and G and the subcontractor's SSAHP for Site I did not contain detailed discussions of specific work tasks to be performed by employees or the hazards associated with those tasks. As a result, SOPs associated with each job or task had not been prepared.

The contractor's SSAHP for Site C contained task- and operation-specific hazard analyses and safety and health procedures that covered general operations, but they were not specific enough for prescribing control methods and PPE for each job task. The plan identified some safety hazards for each operation but did not describe SOPs for protecting employees from these hazards. The SSAHP did, however, contain personnel and equipment decontamination procedures used at the site.

The SSAHP for Site A contained procedures and practices that did not reflect actual practices employed by workers onsite. Examples included PPE requirements, decontamination procedures, and work zone identification procedures.

SOPs should be developed as work progresses. Plan as we may, field adjustments are a part of life at most sites. As these situations unfold, JSAs, JHSa, and SOPs must be reviewed, redeveloped, adjusted, or developed. These items should become part of the SSHAP and reviewed with workers on an ongoing basis. We recommend that JHAs or their equivalents be used as a training aid to new site hires or for workers who get transferred into a different work area. These JHAs should be reviewed on a regular basis to ensure accuracy and applicability.

E. Emergency response elements of the safety and health program must be fully implemented as described in the program. The emergency response plan for a hazardous waste site is one of the key components of a site's SSAHP. Major elements of the emergency response plan include coordination with local organizations that provide emergency response services (i.e., fire department, health care facility, and local emergency response center), training employees in emergency response alarms and evacuation procedures, and conducting drills to determine the effectiveness of the emergency response plan. Requirements for developing and implementing emergency response plans are found in paragraph (l) of the HAZWOPER standard. The audit teams found that none of the sites had established comprehensive emergency response procedures consistent with the HAZWOPER requirements. The audit team at Site H did, however, conclude that an effective emergency response program was in place.

Deficiencies common to most sites were inaccurate emergency contact lists and a lack of communication with local emergency response organizations. Only Sites C and H, and one of the contractors at Site J had accurate site emergency contact lists and had contacted local emergency response organizations. The contractor at Site A had an accurate list of responders who were aware of the site location but unaware of the hazards associated with PCBs, the main contaminant at the site. At Site H, local emergency response organizations indicated that they were familiar with the site and its emergency response requirements.

Additionally, the Site H contractor paid for site-specific emergency response training consistent with the requirements of 1910.120(q) for members of the local fire department so that fire department personnel could respond to emergencies within the site's exclusion zone. This contractor also provided the fire department with weekly project status reports to inform firefighters of the location and nature of current site activities. One contractor at Site J provided the local hospital with necessary information about site hazards, and the hospital developed written procedures for treating potentially contaminated accident victims. The fire department near Site J, however, had been using MSDSs from 1985 as a characterization of site hazards, which indicated a failure to coordinate the site emergency response plan with local, state, and federal agencies.

Notes of a meeting attended by local fire departments, EPA, and Site J contractors, however, indicated that fire prevention, firefighting procedures, and potential hazardous exposures occurring as the result of fire were discussed and that local fire response teams and other site contractors would be made aware of current hazards.

In contrast, contractors' emergency contact lists for Sites B, E, I, K, and the remaining contractors at Site J were inaccurate, and not all local emergency response organizations identified on the lists had been informed about the sites. This lack of coordination was particularly troublesome for Site E which is located in a rural area with no 911 service and relies largely on volunteer emergency response organizations. At Site I, the subcontractor had not informed personnel at the only area hospital with a burn unit about site hazards and potential emergency medical needs that could arise from the use of high-voltage power lines. In addition, although the Site J contractor discussed above had contacted the local fire department and medical center, the local hospital did not know about the presence of all of the Site J contractors.

It appears that the deficiencies outlined in the preceding section do not indicate a plan that was lacking but instead a lack of execution. The underlying factor as to why this lack of execution exists could be due to a less than desirable safety culture.

Another deficiency common to many of the sites inspected was inadequate emergency response training. For example, at Site K, employees indicated that they had not been trained to use fire extinguishers, even though the written plan and site hazard communication training suggested that employees could be called upon to respond to small fires that could be controlled with onsite extinguishers. Similarly, the Site F plan did not contain a provision that all employees expected to use portable fire extinguishers must receive training in fire extinguisher use.

At Site E, personnel had not been trained in how to shut off the LPG tank in case of an LPG line rupture or leak, even though a potential rupture or leak was discussed as a hazard in the SSAHP.

Interviews at Site I indicated that emergency response planning and training had been poorly implemented and that training in emergency spill containment and fire extinguisher use had not been provided.

All contractors except for those at Sites C and H also had major deficiencies in their provisions for conducting emergency response drills. Site H was the only site at which the contractor conducted confined space rescue and emergency evacuation drills, as required by 1910.146(k)(1)(iii) and 1910.120(1)(3)(iv). In addition, the Site C SSAHP described procedures for employees to follow in the event of a fire, explosion, or equipment failure and contained procedures for using hand signals and emergency alarms.

In contrast, emergency evacuation procedures had not been rehearsed at Sites E and K, and some employees at the Site K expressed doubt that the employee alarm system (portable air horns) would be audible above site noise levels or would be accessible for all emergencies.

The SSAHP at Sites B, G, I, and J did not establish provisions for emergency response rehearsals or drills. The Site G and I SSAHPs did not describe procedures for testing and maintaining alarm systems, while the SSAHP at Site F did not contain site-specific emergency alerting procedures, including the exact type and location of alarm systems. The written emergency response plan for Site E did not indicate the frequency of rehearsals, the location of evacuation alarms, the procedure to evacuate when those alarms went off, alternate evacuation routes, the radio channel to be used in emergencies, the location of assembly areas for onsite and offsite evacuation, or procedures for testing the alarm to determine its proper operation. In addition, the evacuation alarm in use at Site E was located only in the safety and health office and had not been tested. One Site J contractor did conduct emergency evacuation drills periodically throughout the year so that all employees had the opportunity to participate at least once annually; however, some employees stated that they had never participated in such drills.

The emergency plan at Site C did not contain provisions for conducting periodic rehearsals or evacuation drills, and it was not clear whether site personnel had opportunities to rehearse emergency response situations with outside community organizations.

Site H was the only site at which the contractor had clearly established personnel roles and responsibilities during emergency response activities. At Site H, the prime contractor designated an onsite emergency response team made up of members who were trained in and responsible for confined space rescue and site evacuation assistance procedures. At least four emergency response team members were on-site during each shift and were distinguished from other personnel by green hardhats. The Site H prime contractor limited emergency response actions for most employees to spill containment activities. These employees were trained to evacuate the site in the event of other site emergencies or potential emergencies, and audit team interviews indicated that the employees were familiar with evacuation routes and procedures.

At Site K, rescue and medical duties for the site had not been planned and specified, and management personnel and employees did not have a thorough understanding of their respective roles and responsibilities during site emergencies. The Site B SSAHP mentioned that employees would be briefed on emergency evacuation routes, potential site emergencies, and employee roles and responsibilities; however, the plan provided no site-specific details describing these emergency response procedures. The SSAHPs at Sites F and I lacked clear and consistent descriptions of personnel roles, lines of authority, and methods of communication during emergencies.

For example, one section of the contractor's SSAHP for Site F required personnel to evacuate the site during an emergency, while other sections of the plan indicated that personnel may respond to spills, leaks, or fires. Neither contractors' SSAHPs at Sites C or G identified the individuals responsible for coordinating emergency response activities.

Many of the sites did not effectively identify the nature and location of potential emergencies. The contractors' SSAHPs at Sites A, B, F, and K did not include site-specific information about the nature and source of potential emergencies. The plan in effect at Site H did not provide a description of the nature and location of potential spill hazards and emergencies, nor did it indicate the type of spill containment equipment available or the locations on site where this and other emergency response equipment was stored.

At Site I, hazards associated with the thermal oxidation unit had not been discussed in site-specific training. In addition, the Site I subcontractor's SSAHP lacked a description of the types of potential emergencies associated with site operations.

Other problems with the sites' emergency response procedures stemmed from deficiencies in the layout and content of the emergency information in the SSAHP. The contractors' SSAHPs at Sites F, G, and I did not include a site map illustrating emergency evacuation routes and designated rally points. The contractor at Site G, however, claimed that this information is disseminated to employees at safety briefings. The Site G SSAHP explained the basic equipment and procedures needed for emergency response but lacked important detail needed for successful implementation of the program. For example, the equipment list did not specify the number and locations of fire extinguishers. The plan called for emergency equipment to be available at all active work areas but did not identify specific locations. The Site G contractor's SSAHP referred to a separate contingency plan, which apparently described much of the information missing in this chapter, but did not state where this supplemental plan was kept on site. Similarly, the contractor's SSAHP at Site I did not contain a description or the location of site emergency response equipment and PPE. The emergency response plans of both Site K contractors were scattered throughout their respective SSAHP documents rather than being consolidated in a separate section of the document as required. One specific concern that the audit team had at the Site J was the lack of emergency coordination between the prime contractor and subcontractors. Site J's prime contractor used an emergency contingency plan, prepared to comply with 40 CFR 264.50, Subpart D.

The 40 CFR regulations have more limited worker safety and health requirements than 1910.120, which resulted in certain deficiencies in the emergency response plan. Subcontractor employees indicated uncertainty about correct evacuation procedures and stated that they were unaware that such a plan was currently in place. Due to the departure of the safety and health manager at this site, employees did not know who to call in an emergency or where the number would be listed.

The requirements regarding emergencies are well documented. Obviously, emergency phone numbers should be checked for accuracy and completeness. A particular person, as specified in the SSHAP, should drive the route to the hospital and note any changes to the hospital route map. The SSHAP must contain emergency requirements and discuss the requirements in a site-specific manner. If an on-site meeting with the local emergency responders is a requirement, this must be included in the plan. Notes should be taken during this meeting, and the local emergency responders should be kept informed if any significant changes have occurred.

Notice that the findings not only pointed out deficiencies in the plan, but difficulties with execution. This lack of execution may stem from a less than desirable safety culture.

F. All site control elements of the safety and health program must be fully implemented as described in the program. The purpose of site control requirements is to ensure that only properly trained and authorized individuals enter those areas of the site with potential hazards, and that, in the event of an emergency, rapid assistance can be rendered to employees working in the exclusion zone. This section discusses the findings of two components of site control: the establishment and maintenance of site work zones and the establishment and implementation of appropriate confined space procedures.

Site Work Zones

One common deficiency in the sites reviewed was the lack of an accurate, up-to-date site work zone map. Of the sites reviewed, only the Site H contractor had established site work zones that were clearly marked on a site zone map. The SSAHP for Site G contained a general discussion of the types of work zones established at the site and the kinds of activities that took place within each zone; although the SSAHP claimed to contain a site map, the map was not available for review. The contractor's SSAHP at Site C contained one map that covered the entire site area, but it did not contain more detailed maps showing locations of support areas, exclusion zone boundaries, or decontamination facilities. The work zone maps for Sites B and D did not accurately reflect actual site work zones, and SSAHPs provided by contractors at Sites I and K did not even contain site maps showing the location of work zones.

Site H was the only site at which the contractor had implemented comprehensive and effective site control elements. The Site H contractor had established site work zones, a buddy system, and site communication procedures consistent with 1910.120(d). This contractor had also established exclusion zones and contamination reduction zones to control migration of site contaminants to clean areas of the site when work within these areas introduced the potential for exposure to hazardous contaminants. The audit team supported this contractor's use of flexible and temporary work zone boundaries based on monitoring results and hazard determinations.

The contractor's site control elements at Site B were not comprehensive; however, the contractor's SSAHP did address site entry and training requirements and mandated that all personnel, including subcontractors and visitors entering the exclusion zone or decontamination zone, meet HAZWOPER training requirements.

Forty-hour training was required for personnel entering the exclusion zone, and additional supervisory training was required for site supervisors. Site control procedures described in the Site C contractor's SSAHP included maintenance of site control logs at each access point, use of red tape or chainlink fencing to demarcate hot zones, and use of the "buddy" system in all exclusion and contamination reduction zone areas. Site communications relied almost exclusively on visual sighting of employees; the plan did not describe the use of two-way radios. This suggested that all employees in hot zones can be observed continuously from the support zones.

Significant deficiencies in site control procedures existed at Site K. For example, the Site K subcontractor had not established a contamination reduction zone (CRZ), to physically separate the support zone from the exclusion zone, as required in the site plan.

The bench where decontamination took place was only a few feet from the thermal unit and was not isolated from exclusion zone activities; employees moved freely between their work stations in the exclusion zone and the decontamination bench. In addition, the subcontractor did not conduct monitoring activities to support work zone designations. Neither Site K contractor used warning signs to delineate exclusion zones, as required by their respective health and safety plans.

At the time of the Site I audit, only two established work zones remained on site: an exclusion zone encompassing the waste treatment area and a clean zone encompassing the remainder of the site. According to the contractor's project manager, EPA approved this reduction in site work zones. As a result of this zone designation, workers exit the exclusion zone directly into a clean zone, removing and discarding PPE in a barrel adjacent to the operations trailer.

The audit team took wipe samples from the surface of the discarded PPE and analyzed them for metals, pesticides, and SVOCs, but found no detectable contamination. The Site I contractor, however, did not have additional sampling data from different days or varying circumstances to verify that on a consistent basis, contamination was not being spread to clean areas of the site because of the lack of decontamination operations.

The Site E contractor had established fixed work zones based on the potential for exposure but adjusted the barriers to accommodate certain activities, such as thermal unit maintenance. While adjusting work zone boundaries according to the potential for contamination is acceptable, worker behavior indicated confusion about the zone boundaries and associated work practices and PPE requirements. In addition, the audit team observed Site E workers leaving the exclusion zone without performing required decontamination procedures.

All contractors at Site J had deficiencies in their work zone practices. The contractors had established work zones based on the potential for exposure associated with specific work tasks.

Temporary exclusion zones, demarcated by red tape, were established whenever maintenance tasks increased the potential for exposure to hazardous contaminants. These zones were removed once visual inspection by a safety technician indicated that the area was clean.

The boundary between the clean support zone and the potential exclusion zone was not clearly demarcated. Employees were told that the boundary was an invisible line drawn between a nearby telephone pole and the corner of a drum storage building. Another problem was that the access road used to travel between the administrative offices and the support zone was used by another contractor to transport drums to the drum-handling building. Even though this situation presented potential contamination issues, vehicles used to transport personnel were not routinely decontaminated, thus increasing the possibility of contaminating the administrative offices. Wipe samples confirmed this concern. Also, drums and other debris were seen on the site of a small abandoned paint factory located near the access road between the administration trailer and the support zone. Although potential hazards had not been characterized, the contractor still instructed employees not to go near the paint factory. This area, however, was identified as the most likely place for a drum to spill during transport, and emergency spill containment equipment was kept by the side of the road near the abandoned factory. This situation may represent a violation of the site characterization requirements of 1910.120(c), even though the prime contractor made it clear that it had not been given authorization to investigate whether the facility presented a potential hazard to its employees.

Another Site J contractor had no CRZ through which contaminated material was to be transported from the exclusion zone to the clean support zone. Additionally, this contractor did not consistently use the buddy system. For another contractor on Site J, the site was loosely zoned and characterized. Employees could generally indicate what areas were "hot," although they were not certain of specific hazards. Some employees were more concerned with the stability of site structures with sagging roofs and broken doors than with chemical hazards. The written description for the exclusion zone in one Site J contractor's SSAHP appeared inconsistent with the actual zone designations. According to this SSAHP, CRZs would be defined on a case-by-case basis, but in practice the entire northern parcel appeared to lack CRZs for personnel and equipment that met the criteria described in the Four-Agency Manual, EPA guidance documents, and other industry literature.

Site D lacked a sufficient CRZ and also lacked access/egress control for the exclusion zone. The site control plan did not accurately identify the function of the CRZ as a buffer zone between the exclusion zone and the support zone, and there was no buffer area between the decontamination pad and the road that runs adjacent to the pad, marked as a support zone. Also, an exclusion zone log-in procedure for tracking personnel who enter and exit this zone was not used on site as called for in the SSAHP.

The primary contractor at Site A had identified clean zones, buffer zones, and related site control procedures in its written plan; however, onsite implementation differed from those specified in the plan. For example, the exclusion zones identified in the plan at the upper pad of the wastewater treatment plant, the dredge area, and the solid waste storage area were marked with signs requiring PPE, but were not labeled with red banners as called for in the plan. In addition, the exclusion zones did not have controlled access through one point of entry as described in the plan, nor were the buffer zones established and demarcated with yellow banners.

The results of the audits indicate that the delineation of zones is easy to put into a plan but difficult to keep current. The nature of remedial work demands flexibility. As sites become remediated, the exclusion zone boundaries change. This is not a situation that is easily handled in a plan, but should be reviewed as other site documents on a predetermined regular basis.

Confined Space

The contractors' SSAHPs at Sites B, F, and K had general confined space provisions but lacked site-specific confined space procedures. For example, SSAHPs for the Site K contractor and subcontractor had written confined space entry programs, but the programs did not establish site-specific rescue procedures or identify the confined space hazards present on the site. The job hazard analyses for both programs failed to address site maintenance tasks that could involve confined space entry and hot work hazards. The programs also failed to identify the specific person or position responsible for supervising confined space entry procedures and the location of permit-required confined spaces on site. Interviews with both contractors indicated confusion about rescue procedures.

Employees stated that they had received confined space training and were prepared to perform confined space rescue, but they had not rehearsed rescue procedures. In addition, the confined space entry permit form for both contractors did not ask for all required information. For example, the form did not require documentation of the duration for the permit, the intended communications procedures for entry operations, or documentation of hot work performed during confined space entry operations. Completed permits did not contain documentation of hot work performed during confined space entry operations, even though hot work had been performed during such operations at least twice during the project.

Similarly, the Site B contractor's SSAHP provided corporate policy and procedures for permit-required confined space entry but lacked the site-specific detail necessary to describe the application of the corporate policy to procedures at the site. For example, the SSAHP did not identify specific components of the thermal treatment unit that presented confined space hazards, nor did it describe the specific circumstances or procedures that would require employee entry into these areas. In addition, the plan stated that the contractor would maintain an onsite employee confined space rescue team, but did not identify the members of this team. The SSAHP for Site F also contained a generic confined space entry program but did not identify the specific location of confined space hazards present at the site.

Contractors at Sites E, H, and J had documented confined space programs but had not fully implemented these programs. The Site H contractor had established a permit-required confined space entry program consistent with HAZWOPER requirements; however, onsite procedures were not completely consistent with the written program or OSHA requirements. For example, the confined space permit form used at Site H was not the form included in the written program. The audit team also found evidence that employee training was insufficient for safe performance of assigned confined space entry duties. The audit team's review of canceled permits at Site H indicated that site personnel occasionally failed to record oxygen levels and other measured atmospheric concentrations as required by site SOPs. At Site E, the contractor's confined space permits had been developed and were required in the SSAHP, but may not have always been completed in actual practice. Names of authorized entrants and standby personnel and the identification of required PPE were not recorded on the entry permits. At Site J, the contractor's buddy system and confined space procedures were in place, although interviews suggested that employees assumed they would be responsible for confined space rescue and were unaware of requirements in 1910.146(k) to rehearse related rescue procedures.

The contractor at Site A had a written confined space program; however, the permits used onsite were different from those specified in the plan. In addition, confined spaces onsite were not labeled, rescue drills had not been conducted, and employee training records were not available onsite.

During the development of the SSHAP, items such as the specific names of the rescue team might not be known. Or, over time, the rescue team members may change. The specific members of these teams should be updated on a regular basis. This basis should be specified in the plan. As stated in the preceding section, the plan needs to be regularly reviewed and updated as necessary.

G. The safety and health program must include procedures for monitoring the effectiveness of PPE, decontamination procedures, and housekeeping programs. Safety and health supervisors at hazardous waste sites need to evaluate the effectiveness of their safety and health programs on an ongoing basis to ensure that the established SOPs are appropriate. Monitoring the effectiveness of these programs is required under paragraph (b)(4)(iv) of the HAZWOPER standard. In general, audits discovered that safety and health personnel have not established objective procedures for monitoring the effectiveness of certain elements of their programs, in particular the use of PPE, decontamination procedures, and housekeeping procedures. The effectiveness of these program elements can be assessed in a variety of ways, such as the collection of wipe samples on decontaminated equipment and on surfaces in clean areas, analyzing the final decontamination rinse water for the presence of contaminants, or visual inspection of PPE for signs of leakage or failure.

Contractors at Sites C and H had acceptable programs for monitoring the effectiveness of PPE and decontamination procedures. The Site H contractor had established PPE and decontamination programs consistent with HAZWOPER requirements. The Site C contractor developed decontamination procedures that included specifications for periodically evaluating the effectiveness of decontamination methods through the performance of leak testing for PPE, visual observation of used PPE, wipe sampling of protective clothing surfaces, and chemical analysis of cleaning solutions. The contractor's PPE program at Site C included procedures for assessing the effectiveness of PPE, such as testing for leaks, visually observing PPE during use for signs of contamination, and recording actual and suspected PPE problems in the daily site log. In addition, the respiratory protection program that is part of the PPE program mandated the use of qualitative or quantitative fit testing at the time a respirator is assigned and semiannually thereafter.

Responsibility for respirator cleaning, inspection, and maintenance rests with each employee; the safety and health officer determines whether employees are maintaining their respirators properly or if additional employee training is required. Many other sites did not have effective monitoring programs in place. Neither of the Site K contractor or subcontractor SSAHPs included provisions for evaluating the effectiveness of decontamination procedures. According to these documents, the adequacy of equipment decontamination was determined by visual inspection alone. The equipment and personnel decontamination procedures of Site K's prime contractor were acceptable, but deficiencies existed in the subcontractor's procedures. The level of PPE used at Site K was not based upon site- and task-specific hazards, and the use of chemical protective clothing was not supported by measurements of surface contamination on the clothing. For example, for both contractors, existing monitoring data did not support the need for full-face respirators, making the associated limitations in worker communication, peripheral vision, and respiration unnecessary.

At Site I, personnel and equipment decontamination procedures were not monitored for their effectiveness in accordance with HAZWOPER requirements. The Site I subcontractor did not have provisions for particulate sampling, evaluating exposure to pesticides and herbicides, or evaluating the effectiveness of site zone boundaries and personnel decontamination procedures. Additionally, monitoring had not been conducted to verify that decontamination was not necessary for employees who leave the exclusion zone and enter a clean zone without undergoing decontamination.

The SSAHP developed by the Site G contractor did not indicate that the contractors routinely conducted job- or task-specific hazard analyses. In addition, the SSAHP did not specify that PPE selection for jobs and tasks must be based on the analysis of the health hazards associated with each job. Furthermore, the SSAHP contained no procedures for objectively determining the effectiveness of decontamination of personnel or equipment. The decontamination program required incineration of all materials that could not be readily decontaminated; such materials were placed in labeled disposal containers. The program, however, did not address storage of these materials until such time that the incinerator was operating.

Other than wipe sampling of clean areas, the Site J contractor did not implement procedures to evaluate the effectiveness of personal decontamination methods. Decontaminated equipment was tested by wipe sampling. Further decontamination was performed as necessary until wipe samples fell below the contractor's trigger levels.

Contractors at Sites B, D, E, and F had not evaluated the effectiveness of their PPE programs and did not monitor decontamination procedures through the use of surface sampling or other quantitative means. At Site E, improper use of PPE was observed several times in the exclusion zone, and monitoring data was not kept current in the computer when the SSO was off site for an extended period of time.

The contractor at Site A had no process in place to evaluate the effectiveness of its decontamination procedures.

Assessment of the effectiveness of PPE is an important part of any SSHAP. Writing an appropriate plan is challenging, but it is typically less challenging than the execution of safe and healthful work activity. What appears to be lacking in the preceding scenario is execution. There are a variety of reasons why a well-written plan might lead to a lack of execution. Although writing an appropriate plan is challenging, it is truly the execution of the plan which is most difficult. Again, keep in mind that the authors believe that planning provides the basis needed to achieve a safe and healthful work place. An effective SAHP is a basic step, but no plan, in itself, can overcome poor execution.

H. Self-audit site inspection and abatement tracking programs must be formalized and effectively implemented.

The overall effectiveness of the safety and health program must be evaluated, in part, by conducting regular inspections and audits. In addition, the SSAHP should include a mechanism for following up on corrective actions recommended by the site safety and health officer during safety inspections. All hazard abatement actions identified by the site safety and health officer should be tracked to ensure that the corrective actions have been implemented and the hazard(s) have been controlled. The program should designate individuals to periodically inspect work areas and ensure that hazard abatement has been accomplished. Paragraph (b)(4)(iv) of the HAZWOPER standard contains the requirement that the site safety and health supervisor, or a knowledgeable designee, perform periodic inspections to evaluate the effectiveness of the program.

Site H was the only site for which the contractor had developed and effectively implemented inspection procedures consistent with the HAZWOPER requirements. Site safety technicians conducted daily and weekly inspections during each work shift. The technicians recorded deficiencies on an inspection checklist, and the site health and safety officer

(HSO) and project manager reviewed and signed each checklist. The prime contractor then compiled a weekly "Observations Report" from the daily inspection records, which included a list of noted deficiencies and the necessary corrective actions. The deficiencies appearing on this report were carried forward to following weeks until corrective action was taken. The HSO also conducted a monthly safety and health inspection. A review of site records indicated that, in most cases, the contractor had implemented corrective actions in a timely fashion. In addition, the site walk-through suggested that site safety and health practices and procedures were generally effective.

The contractors' SSAHPs at Sites I and K required that safety and health program inspections be conducted; however, these requirements were not effectively implemented at either of these sites. Both Site K contractors required the HSO to conduct daily inspections, and both stated in their written plans that hazards would be immediately corrected. Neither contractor, however, had established hazard abatement procedures to ensure the prompt correction of hazards, and site records for both contractors indicated that hazard abatement activities were either not documented or not completed. For example, the subcontractor's daily safety log contained several notations of safety hazards, including an unstable concrete well and storage of diesel cans near the propane tank; however, later log entries and site records did not track the abatement of these hazards.

At Site I, the prime contractor's SSAHP required daily site inspections, the documentation of safety and health deficiencies, and the abatement of deficiencies. Records of site deficiencies, however, were kept intermittently, and hazard abatement was not documented. The subcontractor's SSAHP did not address site inspections and hazard abatement, but its TSCA permit application included requirements for site inspections. Inspection documentation, however, was not available on site, and the site manager was unaware of these written requirements. The site manager did, however, indicate that he conducted site inspections using a mental checklist and that he conducted inspections of remediation equipment before each use.

The SSAHPs for contractors at Sites E, F, and G did not establish specific requirements for self-audits to identify and correct any deficiencies in the effectiveness of the plan. The Site F contractor's SSAHP stated that the health and safety manager is responsible for continued evaluation of the plan's effectiveness, but the SSAHP did not establish specific requirements. At Site E, the contractor used chronological notebooks to file all daily logs; however, health and safety program elements or related activities were not properly identified in the logs, making it difficult to verify compliance with the SSAHP or OSHA requirements or to determine whether recognized hazards had been abated.

Again, writing a plan that calls for regular inspections appears to have been implemented by the sites audited. What has been lacking appears to be the execution and recordkeeping.

Procedures to monitor and reduce heat stress need to be effective. Perhaps the greatest health hazard facing hazardous waste site workers is heat stress, exacerbated by the use of impermeable chemical protective clothing. Ideally, a comprehensive heat stress program will contain several elements, including environmental and medical monitoring (i.e., measurements of pulse rate, oral temperature, and/or weight loss), issuance of heat alerts, implementation of work-rest regimens when site conditions warrant, provisions for fluid intake and shaded rest areas, and regular training of employees in recognizing the signs and symptoms of heat stress in themselves and their fellow workers.

For the most part, contractors at Sites C and H had established effective heat stress monitoring programs. The SSAHP for Site C contained detailed procedures designed to protect workers from heat and cold stress. These procedures included environmental sampling and medical monitoring of workers when ambient temperatures reach 70 degrees F, and heart rate and oral temperature monitoring at the beginning of each worker's rest period when ambient temperatures reach 80 degrees F. The contractor at Site C modified work and rest schedules based on the results of medical surveillance and monitored weight loss during hot periods to ensure employees maintained sufficient fluid intake. In addition, this contractor's SSAHP contained a comprehensive discussion of the signs, symptoms, and treatment of heat- and cold-related disorders.

The contractor at Site H had established SOPs for site hot work consistent with the HAZWOPER requirements. This contractor required initial body weight and pulse measurements and core temperature readings at fixed intervals during the work shift as well as exit body weight and pulse measurements. Workers were directed to stop work if their core temperature exceeded 100.4 degrees F or if they felt uncomfortable. Site records indicated that heat stress monitoring was conducted on a regular basis when the ambient air temperature reached or exceeded 70 degrees F.

Contractors at Sites E, F, G, and K referred to and/or performed some heat stress monitoring procedures, but none had a comprehensive heat stress plan. Although an effective heat stress program, based on NIOSH recommendations, was included in the SSAHP for Site E, the procedures used on site varied from the program outlined in the SSAHP. Employee body temperature measuring and monitoring was not conducted as required by the SSAHP. The employee work/rest schedule was not actively monitored by the SSO to assure that the heat stress prevention plan was being followed. Employees monitored their own pulse and blood pressure prior to dressing out and before leaving decontamination areas, and they filled out log sheets that were maintained in the decontamination trailer. The lack of attention to heat stress monitoring, however, was evident in these daily logs.

For example, for a caustic spill response performed by employees on site, the log indicated that one worker experienced heat exhaustion during the cleanup and was absent from work the next day, "likely due to heat exhaustion from the caustic spill." No other entries in the log discussed the use of heat stress monitoring or prevention practices, suggesting that such practices were not always implemented on the site.

The contractor's heat stress program at Site G called for measuring ambient temperature, instituting necessary controls, providing rest areas, and establishing work/rest schedules. The program, however, provided no details about when controls were to be implemented, what controls were to be implemented, when work/rest schedules were to be used, how to determine the appropriate schedule, or when to conduct medical monitoring for heat/cold stress. The plan referred to an appendix that contained a general corporate program for heat stress prevention. The corporate program mandated the development of a written procedure for operating groups, but this was not included in the SSAHP.

The heat stress plan at Site F detailed methods for monitoring workers' heart rate and oral temperature, but did not designate the personnel responsible for performing such monitoring, nor did it include information about the availability or location of instruments for actually monitoring such parameters. In addition, the plan did not identify or discuss the location and availability of drinking water.

The heat and cold stress program in the contractor's SSAHP at Site B appeared to be a statement of corporate policy and contained no details about site-specific heat stress or cold stress program procedures at the site. The SSAHP for Site J did not appear to have established heat stress SOPs, but indicated that workers should evaluate how they were feeling. The SSAHPs for Sites D and I had no discussion of heat stress.

The contractor SSAHP at Site A provided for heat stress monitoring to begin when the temperature rose above 70 degrees F. The OSC at this site indicated that this rarely happens during the summer months, and, thus, heat stress monitoring had not been conducted.

The audit indicated that heat stress programs were generally included in most sites that were audited. The difficult part again appears to be in execution of the SSHAP, although, in some cases, no provisions were included in the plans for heat/cold stress.

K. Employers must develop and implement training programs to inform workers of the degree of exposure they are likely to encounter and how they should avoid adverse situations.

Employers are required to develop and implement a program to inform workers performing hazardous waste operations of the level and degree of exposure they are likely to encounter. This information needs to be contained in the SSAHP. Employers are also required to develop and implement procedures that introduce workers to the most effective technologies that provide protection in hazardous waste operations. The employer must develop a training program for all employees exposed to safety and health hazards during hazardous waste operations. The training program should educate supervisors and workers to recognize hazards and prevent exposure to the hazards; to properly use and care for respirators and other personal protective equipment; to understand engineering controls and their use; to use proper decontamination procedures; and to understand the emergency response plan, medical surveillance requirements, confined space entry procedures, spill containment program, and any appropriate work practices. With minor exceptions, contractors at Sites A, C, H, and K had implemented training programs that were generally consistent with the HAZWOPER requirements. Contractor records at Site H indicated that training was not only generally consistent with the HAZWOPER requirements, but was also well documented. At Site A, current 40-hour and 8-hour refresher certificates were available for all onsite workers; however, it was not clear when employees had received three days of supervised field experience as required by the HAZWOPER standard. Additionally, daily safety meetings were conducted at Site A but not documented. At Site K, both contractors appeared to provide training in accordance with HAZWOPER requirements. Records for both contractors indicated that site workers had received 40-hour initial training and 8-hour annual refresher training as appropriate, and project managers for both contractors had received supervisory training. In addition, the prime contractor maintained training records for other subcontractors used on the project, and the SSAHP required daily tailgate safety meetings. The contractor's SSAHP at Site C described the kind and amount of training required for four groups of employees: onsite supervisory personnel, general site workers, workers on site less than 30 days and not likely to be exposed above permissible or other published exposure limits, and nonexposed workers. The training specified for each of these groups was in accordance with HAZWOPER requirements; however, the SSAHP did not specify who provides the training or how its adequacy is verified.

In addition to the required training, the SSAHP stated that all employees must participate in tailgate safety meetings at least once per week or before starting a new job or work task.

Other sites provided general provisions for acceptable training programs but lacked the site-specific detail necessary to implement a successful program. For example, the contractor's plan at Site F contained a requirement that all project field personnel receive training in accordance with applicable OSHA standards, including a minimum of 40 hours of hazardous waste operations training. The plan did not contain, however, the details of required site-specific training, such as the safe use of engineering controls and equipment on site. The contractor's SSAHP training program at Site G did not address how employees would be provided with three days of supervised field training as required under the HAZWOPER standard. Instead, site-specific training was provided in a briefing.

Some sites were not providing all training necessary. Contractor training records at Site E indicated, with minor exceptions, that 40-hour HAZWOPER training/refresher training was current for all workers; however, supervisor training was not provided by the corporate office, and only one worker, an operations engineer who did not supervise others, had received supervisor training. In addition, the contractor produced no evidence that line supervisors had received at least 8 hours of supplemental supervisor training for hazardous waste supervisors. Contractor training records at Site J showed gaps of several years between initial and refresher training for some employees and no initial training documentation for one employee. Not all employees who performed supervisory duties were documented to have had supervisor training.

The contractor's SSAHP at Site B mandated that all personnel, including subcontractors and visitors, entering the exclusion or decontamination zone meet HAZWOPER training requirements. Forty-hour training was required for personnel entering the exclusion zone, and additional supervisory training was required for site supervisors. The SSAHP also required that documentation of three-day supervised field experience be submitted to the SSO along with other training documents, but the plan did not address how the contractor would provide supervised field experience for employees who had not received such experience or did not have the appropriate certification.

Training was found to be deficient in a variety of areas. It appears that for the most part, the plans were adequately written, but the execution was lacking.

L. A medical surveillance program must be in place to assess and monitor the health and fitness of employees. A medical surveillance program helps assess and monitor the health and fitness of employees working with hazardous substances. The contractors at Sites A, E, H, and K and one subcontractor at Site I appear to have established medical surveillance programs that with minor exceptions were consistent with HAZWOPER requirements.

Contractor records at Site H indicated that medical tests and procedures included annual examinations that addressed site-specific hazards and were provided with the frequency required by the standard. Records at Site K showed that employees of both contractors had received recent comprehensive medical examinations, and copies of the physician's written opinion were maintained for each employee. These medical records, however, contained no documentation of termination examinations. The project manager at Site K told the audit team that termination exams were made available to employees. At Site I, employee interviews indicated that the subcontractor had implemented a medical surveillance program and that examinations were offered on a schedule consistent with HAZWOPER requirements; however, records were not available for 8 of 14 employees covered by the program. The site manager said that these records were filed at company headquarters.

The contractors at Sites B, C, F, G, I, and J had significant deficiencies in their medical surveillance programs. Medical surveillance practices at Site B were not consistent with the requirements that all employees shall have termination physicals and that the examining physician shall be responsible for determining the need for additional monitoring. The contractor's SSAHP at Site B appeared to require termination physicals for exclusion zone personnel only and allowed the SSO to determine the need for additional monitoring. In addition, the SSAHP did not provide site-specific medical monitoring requirements or the schedule for providing medical exams. The plan did state, however, that all personnel, including subcontractors and visitors, entering the exclusion zone or decontamination zone must have received "appropriate medical monitoring" in accordance with HAZWOPER.

The schedule for providing medical surveillance at Site F did not include the required medical surveillance at termination of employment. At Site J, annual medical exams appeared to have been scheduled and documented appropriately for current employees, but only two termination examinations were documented, despite other known employee turnover discussed in interviews. The contractor's SSAHP at Site C did not describe a site-specific medical surveillance program that reflected site hazards. Sections of the SSAHP dealing with respirator use and heat stress, however, did require employees to be certified for fitness before being assigned to tasks where respirators were required or where heat stress hazards were present. The SSAHP for Site G referenced a corporate medical program that established a recommended content and frequency for medical examinations. The SSAHP, however, did not describe how the content of the medical program related to the employees' jobs, PPE use, health hazards present, or even whether the program was designed based on these considerations, as required in the corporate program. At Site I, the prime contractor had not implemented a medical surveillance program on site, and the project manager was unaware of these HAZWOPER requirements.

The deficiencies revealed during the audit indicated that some of the medical surveillance programs were not site specific. The SSHAP should be written to ensure that workers are tested for not only general physical health, but also for those substances that they might be exposed to during work activities. Besides the lack of specificity, it appears that execution of

the written plan and communication of plan content were lacking in some locations.

V. SUMMARY

The safety and health programs at these sites were generally comprehensive in scope and oriented toward compliance with HAZWOPER and other applicable requirements; however, some of those audited lacked the degree of site-specific detail and trained management necessary for an effective safety and health program. The audits revealed consistent deficiencies attributable to a failure to apply professional judgment appropriately and to pay attention to meaningful details.

These problems were evidenced in several ways:

Hazard analyses failed to consider all the available data describing the safety and health conditions at each site.

Objective measures to evaluate the effectiveness of the site's safety and health program were lacking.

Exposure monitoring programs were targeted toward compliance rather than toward the characterization of employee exposures.

In addition, these audits identified several disincentives and obstacles to altering the safety and health culture at these sites. For example, contractors were often not free to exercise independent judgment because contractual provisions locked them into predetermined activities that did not permit them to respond to changes in site conditions or to new information. At other sites, safety and health officers had the authority to make changes but often did not have sufficient experience in safety and health to properly evaluate situations and impose changes. In summary, OSHA believes that nothing short of a rigorous program of ongoing self-assessment, improved training in hazard recognition and evaluation, enhanced management commitment, and sustained employee involvement in the program will achieve the change in culture needed to move these sites toward excellence in occupational safety and health.

REFERENCES

"U.S. EPA 1984 Standard Operating Safety Guides." Office of Emergency and Remedial Response, Hazardous Response Support Division, Edison, N.J. November 1984.

29 CFR 1910.120, "Hazardous Waste Operations and Emergency Response" (HAZWOPER).

29 CFR 1910.146, "Permit Required Confined Spaces."

American National Standards Institute (ANSI) Recommendation Z117.1-1989, "Safety Requirements of Confined Spaces."

29 CFR 1910.38, "Employee Emergency Plans and Fire Prevention Plans."

Whitfield, P., "EM-40 Hazardous Materials Training Program," memorandum of February 3, 1994.

"Occupational Safety and Health Guidance Manual for Hazardous WasteSite Activities," NIOSH/OSHA/USCG/EPA, October 1985 (Four-Agency Document).

DOE Order 5480.1B, "Environment, Safety, and Health Program for Department of Energy Operations."

DOE Order 5483.1A, "Occupational Safety and Health Program for DOE Contractor Employees at Government-Owned Contractor Operated (GOCO) Facilities."

Appendix B

Choosing a Contractor/Subcontractor

The following information is presented as an aid when choosing a contractor. This information was field tested and proven to be successful over a period of several years. The reader may find that the following information needs to be adjusted in one or more ways to be effective. Number values and limitations certainly vary from organization to organization. We do not believe that the information offered provides hard and fast rules but should be used as guiding general principles.

Any contractor being considered for an award must have a history of performing work in a safe manner. The authors believe that if a contractor has performed work safely in the past, it is likely that the management of the contractor will believe in and practice safe work performance as part of its present and future business philosophy.

Note that the following information will need to be modified to meet the needs of the current organization. The size of the host organization and the roles of management within the organization are key items that need to be considered. This information, as presented, is meant to closely meet the needs of a mid-sized company. But no matter what size company will use this information, it will need modifications. These modifications will include forms and attachments that have not been included. We believe that each host organization should develop its own procedures, including its own forms and attachments to fit its needs, personnel, and business structure.

PURPOSE AND SUMMARY

This procedure provides recommended guidelines to aid in choosing qualified contractor(s) to do work as either a general, prime contractor, or a subcontractor.

No matter how large or small an organization is, there will invariably be many times when a host organization will need to get outside assistance to ensure that certain work tasks will be properly completed.

The process of engaging the required capabilities must be a formal process so that both the organization in need and the contractor/subcontractor are protected in the event of a failure to perform, an accident, or a difference of opinion as to terms or performance. Adherence to the provisions of this procedure will help attain good contracting practices and minimize the potential liabilities to the host organization in contractual relationships.

DEFINITIONS

- **Host organization:** All companies, subsidiaries, affiliates, divisions, groups, joint ventures, or projects of host organization
- **Contractor/subcontractor:** A party in either the prime or subcontractor role, or otherwise providing goods or services to the organization, who performs some of the obligations of a particular prime contract. For example, a contractor hired to put up temporary fencing around a work area to control access would be a subcontractor who must be prequalified.
- **Vendor:** An outside supplier of raw materials, supplies, equipment and minor services needed for the host organization to perform its operations (i.e., Office Max provides office supplies).
- **Flow-down provisions:** Terms and conditions which must be incorporated into a subcontract or purchase order to "pass down" obligations/requirements imposed by the prime contract. These flow-down provisions will provide for certain protections, obligations, or requirements that will modify or add to any work agreement or purchase order terms and conditions.
- **Exclusion/contamination zone:** Project area where contamination of any kind is believed to exist or does exist. Sometimes referred to as a "hot area."

DETERMINING WHO WILL BE PREQUALIFIED

No prequalification is required for vendors who perform simple deliveries or drop ship (i.e., Federal Express, telephone company, bottled water supplier, trash collector, utility companies, supply and material deliveries), and would not involve entrance into an exclusion/contaminated zone.

Proof of insurance is required for vendors who perform warranty work, authorized service representatives, air conditioning service/maintenance, minor maintenance, or minor maintenance/repairs for electrical, plumbing, etc., or landscaping, cleaning, and so on. This excludes individuals who will not exceed $1,500 labor cost and do not require entry into an exclusion/contamination zone.

Prequalification is required for any contractor/subcontractor who:

- Performs work as part of a "contract" or "subcontract"
- Works exclusion/contaminated zones
- Provides rental equipment with operators
- Performs electrical/plumbing installations (as part of a contract)
- Enters any exclusion/contaminated zone
- Performs non-prequalification that exceeds $1,500 per transaction (labor cost)

The vendors described in non-prequalification would not require a work agreement to provide health and safety training or other prequalification requirements. In addition, material cost would be excluded from the not-to-exceed $1,500 amount.

Insurance required under non-prequalification should not be less than $1,000,000. Waivers for a lesser amount of insurance must be cleared through the purchasing manager.

If any particular vendor, contractor/subcontractor or event falls outside of the preceding descriptions, then a judgment must be made by the purchasing manager as to contractor/subcontractor status.

RESPONSIBILITY MATRIX

All Organizations

All company organizations, divisions, groups, or projects must inform Purchasing of the anticipated need for contractor/subcontractor service and provide Purchasing with information regarding the recommended contractor/subcontractor including but not limited to contractor/subcontractor name, principal contact, address, and phone number. Anyone requesting the services of a contractor/subcontractor must provide pertinent information regarding the type of activity or service required.

Purchasing Department

The purchasing department will:

- Act as the interface between contractors/subcontractors and the using organizations or individuals
- Provide potential contractors/subcontractors with documentation requirements and contractor/subcontractor prequalification forms

- Send completed contractor/subcontractor prequalification forms for approval to designated location health and safety professional and QA/QC officers
- Reconcile differences, if required, and establish the contractors/subcontractors on the qualified list or notify them if not approved
- Review the insurance certificates to ensure adequacy of policy limits and policy periods
- Require each subcontractor to submit an updated prequalification form with each new bid more than one year old
- Review changes or exceptions to the work agreement and be responsible for negotiating minor deviations in terms and conditions. Any major changes in indemnity insurance, bonding, consequential damages, warranties or other critical provisions will be reviewed by the appropriate party (i.e., the vice president, purchasing or the legal department, etc). Note: Prime contract flow-down provisions will be incorporated with requests for bid proposals and subcontracts/purchase orders at the time of initiating purchasing activity subsequent to the contractors/subcontractors prequalification procedure. Flow-down provisions will be provided by contract administration to purchasing in a timely manner.
- Maintain qualified contractors/subcontractors letters, attachments, and associated documentation, update the files as appropriate, and prequalify contractors/subcontractors on a yearly basis

Health and Safety Department

The health and safety department will

- Establish the criteria for training the contractor/subcontractor.
- Review the completed contractor/subcontractor prequalification forms and provide an evaluation of the contractor/subcontractor's ability to meet the host organization's health and safety policies and procedures. Contractors/subcontractors must demonstrate their ability to meet established criteria, to the satisfaction of the health and safety professional, in order to be considered for prequalification.
- Ensure that the contractor/subcontractor can implement a comprehensive health and safety program in compliance with applicable regulations, including accident prevention programs, medical surveillance, training, work practice controls, use of personal protective equipment, and so on.
- Audit the contractor/subcontractor prequalification form and program for compliance with this procedure and provide medical surveillance and training policies.

Project Managers or Project Health and Safety Staff

Project managers or health and safety staff are responsible for obtaining and verifying training and medical certifications for individual subcontractors employees assigned to a project.

Subcontractors

The subcontractors must

- Have the ability and willingness to perform work in compliance with host organization policies and procedures, where those procedures go beyond regulatory requirements.
- Submit a written plan describing the hazards and control measures for the work to be done by the contractor, for each new contract or task. This plan must identify (as a specific individual) all competent or qualified persons required by applicable regulations or host organization procedures.
- Provide employee accident experience for the past 5 years, including the current year. The submittal shall specifically include OSHA recordable cases rate, lost and restricted workday cases rate, vehicle accident rate, and number of fatalities with a description of each. The workers' compensation interstate experience modification rate should be less than 1.0, and applicable SIC codes should be noted.

Quality Assurance

The regional quality assurance representatives will review for approval the completed contractor/subcontractor prequalification forms and provide an evaluation of the contractor/subcontractor's ability to meet quality assurance standards.

Contracts Administration

The regional contract administrators (purchasing department or others, depending on the structure of the organization) will provide the prime contractor flow-down provisions for incorporation into solicitation and subcontractors' purchase orders.

In the event of an emergency situation, and a contractor/subcontractor is required in an exclusion/contaminated zone, an insurance certificate naming the host organization as the holder and meeting the requirements of the work agreement must be provided before entry. Afterwards, a complete package must be submitted and approved prior to payment.

PRELIMINARY REQUIREMENTS FOR CONTRACTOR PREQUALIFICATION

General

Under absolutely no circumstances shall the services of a contractor/subcontractor be utilized until the prequalification process is complete. That process involves the completion and return of specific documents transmitted under a letter of request.

The request letter shall enclose a Contractor Prequalification Form; work agreements; Representations and Certifications Form; Certificate of Completion and Release of Lien Form; General Safety Rules Contractors Booklet, and a copy of the General Safety Contractors Receipt Form.

Because the nature of projects and type of exposure vary from one intended use of a contractor/subcontractor to another intended use, a contractor/subcontractor qualified and in good standing with one profit center or project may not be qualified to work for another profit center or project without evaluation based on the characteristics of the new assignment. The purchasing department should review the prequalification files or information and evaluate each contractor/subcontractor with respect to each project to be assigned.

Contractor Prequalification Form

This form documents information related to contractors/subcontractors with particular emphasis on a record of good safety performance and quality which meets your requirements so that the host organization may be provided with safety performance and cost control. Failure to complete and return this form will preclude qualifying a contractor/subcontractor to do work for the host organization.

Work Agreement

The work agreement is the single most important document in engaging a contractor/subcontractor. The carefully composed language of the document covers a host of considerations essential to a proper and legally binding relationship between the host organization and the contractor/subcontractor and should not be altered. However, should some peculiar circumstance related to a particular project dictate, purchasing or a company officer may authorize modification.

If the modification deemed necessary is extensive or fundamental (e.g., relating to such matters as insurance or indemnity obligations), the prior approval of the appropriate person or department (i.e., vice president of purchasing or corporate counsel) will be required.

Purchasing will execute all work agreements. Purchasing will also obtain the financial background or credit reports of unknown contractors/subcontractors and perform reviews of contractors/subcontractors from time to time.

General Safety Rules for Contractors

A copy of the general safety rules accompanied by a receipt form shall be included with the prequalification letter to a prospective contractor/subcontractor. This document sets forth in broad terms the safety requirements with which a contractor/subcontractor is expected to conform while working under contract for the host organization.

Insurance Certificate

A key requirement of the work agreement is an insurance certificate from the contractor/subcontractor evidencing certain levels of coverage, naming the host organization as an additional insured, providing for notice of cancellation, and including a waiver of all rights of subrogation in favor of the host organization. These certificates are normally renewed annually. A comprehensive general and automotive liability endorsement must be a part of the certificate. A contractor/subcontractor may not be considered qualified until a current insurance certificate with acceptable coverage, limits, and endorsements is on file.

Qualified Contractor General File

A file in the name of each qualified contractor/subcontractor shall be maintained in the purchasing office awarding the qualification, with copies as appropriate at satellite or field buying offices. The file shall contain:

- A copy of the prequalification letter request
- The completed Contractor Prequalification Form
- The executed work agreement
- A current insurance certificate with limits as specified in the executed work agreement
- Safety Rules Receipt Form
- Representations and Certifications Form
- Site Safety Rules Receipt Form
- Copies of completed Evaluation Forms from previous projects, if any

GENERAL REQUIREMENTS

Documentation

Ongoing paper work is essential to the proper daily management of a project. Further, that same documentation forms a base of data for certain essential postproject reports and reviews. Therefore, day-to-day documentation of all facets of a project, although sometimes burdensome, is necessary.

Purchase Orders

The purchase order is the means by which the host organization engages a qualified contractor/subcontractor to perform specific services for the host organization. It must contain the following terms in addition to those in the work agreement:

- A description of the work to be performed
- The agreed on time schedule for work accomplishment
- Equipment and facilities to be provided by the host organization (if none, so state)
- Compensation arrangements (amount and payment schedule, i.e., 45 days)
- Special insurance coverage that may be deemed necessary due to the nature of a particular job
- Other special requirements such as medical examinations, safety procedures and equipment, training, quality requirement programs, or other precautions
- Flow-down provisions from the client contract, if any
- Notation should be made on the purchase order, "Subject to terms and conditions of work agreement dated . . ."

The purchase order must be forwarded to the contractor/subcontractor with copies to be placed in the project and purchasing files.

Change Orders

All changes in scope of work will be made by a change order to the original purchase order. If operational circumstances dictate a verbal change in scope, that change must be confirmed in writing to the contractor/subcontractor by a change order as soon thereafter as practical. Copies of each change order will be placed in the project and purchasing files. Invoices from the contractor/subcontractor for payment for work in excess of that specified in the implementing purchase order or any subsequent change orders will not be honored.

Training Requirements

Only personnel who have completed the training prescribed by host organization health and safety personnel shall be allowed to work on host organization projects. Some projects may require contractor certification or special training as specified in the implementing purchase order.

Site Safety Rules

When a contractor/subcontractor is hired to perform work in a potentially hazardous area on one of your facilities or that of a client, the project manager shall provide the contractor/subcontractor a copy of the contractor site safety rules checklist for completion. The form and any other rules specific to that site must be signed, dated, and returned prior to any work being performed at a particular site. An executed copy will be made a part of the project file.

Business Classification and Taxpayer Identification

Contracts with the federal government require compliance with Executive Order Nos. 11625 and 12138 to utilize small and small disadvantaged businesses. The contractor/subcontractor must complete a copy of the representations and certifications providing self-certification of business classification under existing federal definitions. The representations and certifications also provide tax identification information required by the Internal Revenue Service. The original of the representation and certification form should be sent to regional purchasing and a copy included in the project file.

Health and Safety Compliance with Regulatory and Host Organization Requirements

Contractors are required to comply with all applicable federal, state, and local regulatory requirements, in addition to host organization requirements as described in company policies and procedures and the site-specific health and safety plan (H&S plan).

Daily Activity Logs and QA/QC Reports

Contractors/subcontractors shall prepare and submit reports of the work as required. This may include daily technical reports, invoices, or other documents.

Miscellaneous Documents

All correspondence, inspection reports, and other documents pertaining to the project, particularly those validating visits and inspections conducted by outside enforcement agencies, shall be kept in the project file.

Qualification of Subcontractors

All lower tier subcontractors engaged by a contractor/subcontractor providing services to the host organization will be qualified in accordance with this procedure, the same as the contractor/subcontractor, with particular emphasis on the prequalification form, insurance requirements, and safety rules. A qualified contractor/subcontractor file shall be maintained by purchasing for each lower tier subcontractor employed by a contractor/subcontractor.

Project File

A file shall be established and maintained at the job site or home office for each project until the project has been completed. A representative project file should include:

- Signed copies of the work agreement or subcontract and implementing purchase order or contract
- Copies of all change orders
- The receipt for company general safety rules signed by the contractor/subcontractor and any lower tier subcontractors
- Signed copy of any site-specific safety rules
- Signed copies of hot work permits
- Excavation permits (required in California) and excavation records
- Signed copies of confined space entry forms
- Copies of all daily activity logs, qa/qc reports, or other quality-related documents
- Signed copies of all tailgate safety meeting reports
- Records of training conducted by host organization or others

- Copies of work restrictions
- Copies of all bid analysis and award rationale
- Copies of completion certificates which must be received from the contractor/subcontractor prior to final payment
- All other correspondence, inspection reports, approved contractor/subcontractor invoices, and supporting documentation

If a contractor/subcontractor is working on host organization premises or under host organization supervision, typically the project manager is responsible for the establishment and maintenance of the project file. In those rare instances where a contractor/subcontractor is working for the host organization, without direct host organization supervision, the contractor/subcontractor shall be required to maintain the project file. When the project is completed in the field, the project file shall be transferred intact to the contracting host organization office.

Postproject Requirements

A postproject management audit (project audit) is performed to verify that project terms were carried out and that services contracted for were satisfactorily completed. Once these basic objectives have been achieved, it is important that all documents are properly assembled and the project file be closed and stored in a manner permitting rapid reference and retrieval.

Project Reports

All reports delivered to a client, or in the case of in-house projects on host organization premises, the reports presented to host organization management, must be a part of the project file.

Records from Contractors

Any contractor/subcontractor records required by the work agreement, subcontract, contractor, or purchase order that were not collected during the project should be assembled during the postproject phase. Provision for a turnover package containing all reports, drawings, calculations, and other documents required by the contract shall be made. Of particular importance is a listing of all subcontractors that were engaged and copies of a release of lien from each contractor/subcontractor to be made a part of the project file.

Contractor/Subcontractor Performance Evaluation

The performance by the contractor/subcontractor should be noted on an evaluation form and a copy sent to regional purchasing for inclusion in the prequalification file.

Quality Assurance Review

The project review or a quality assurance audit will verify that all required records and drawings are accounted for, filed, and stored as prescribed.

Disqualification of Contractor

A contractor/subcontractor shall be disqualified from providing services to the host organization by failure to conform to any of the requirements of this procedure or to perform satisfactorily on a project. Notification of disqualification shall be circulated by the host organization office to regional purchasing and any other company organizations that might have occasion to engage the services of the contractor.

OTHER REQUIREMENTS

Exception Provisions

All exceptions must have the prior approval of the purchasing department.

Contractor Prequalification Requirements

All prequalifications for health and safety work must be approved by an assigned health and safety professional. A rating system will be assigned to all contractors regardless of work conditions. The rating system is as follows:

A: Fully Qualified

Subcontractors may be used for all hazardous waste activities if they meet the following requirements:

- 40-hour training including 8-hour annual refresher and 3 days on site
- 8-hour supervisor training

- Medical surveillance program
- Active drug and alcohol screening and awareness program
- A written safety program and job-specific safety plan
- Experience modification rate < or = 1
- Written acknowledgment of contractor safety rules

B: Qualified

Subcontractor may perform limited site work (e.g., nonroutine tasks such as surveying, etc.) but may not work in exclusion/contamination reduction zones if he meets the following requirements:

- 24-hour training including 8-hour annual refresher and 3 days on site on-the-job training
- 8-hour supervisor training
- Medical surveillance program
- Active drug and alcohol screening and awareness program
- A written safety program and job-specific safety plan
- Experience modification rate < or = 1
- Written acknowledgment of contractor safety rules

C: Limited Qualification

Subcontractor may be used in support zone or nonhazardous site activities. For limited activities at a hazardous waste site, the scope of work must be reviewed with the health and safety professional before work is started (examples: landscape service, electricians, software development, training, etc.). The subcontractor must meet the following minimum requirements:

- Provide basic safety training to employees
- Experience modification rate < or = 1
- Written acknowledgment of contractor safety rules

D: Qualified for Engineering Design Work

Subcontractor does not have to meet minimum requirements. All work is accomplished in the office. The contractor is not qualified for any fieldwork and cannot be used for any field applications unless the minimum requirements outlined in A, B, or C are met.

E: Unacceptable

Subcontractor does not meet the minimum requirement necessary to perform work and will not be used for any jobs.

NOTE: Contractors unable to meet host organization requirements for accident rates or experience modification rates may submit a written safety enhancement program designed to bring project performance in line with host organization requirements which will be implemented for all work done for the host organization. If approved by the host organization health and safety professional, the safety enhancement plan will become part of the contractor's job-specific safety plan and the contractor may be approved.

Appendix C

Process Safety Management Guidelines for Compliance

A question often asked when dealing with hazardous materials is whether a certain site is compliant with the Process Safety Management Guidelines as well as HAZWOPER guidelines. In order to answer this question, we have modified a Department of Labor document and included it below. Should the reader desire more information on this subject, refer to the OSHA web page for the following publication: U.S. Department of Labor Occupational Safety and Health Administration 1994.
OSHA 3133

PURPOSE

The major objective of process safety management (PSM) of highly hazardous chemicals is to prevent unwanted releases of hazardous chemicals especially into locations that could expose employees and others to serious hazards. An effective process safety management program requires a systematic approach to evaluating the whole chemical process. Using this approach, the process design, process technology, process changes, operational and maintenance activities and procedures, non-routine activities and procedures, emergency preparedness plans and procedures, training programs, and other elements that affect the process are all considered in the evaluation.

APPLICATION

The various lines of defense that have been incorporated into the design and operation of the process to prevent or mitigate the release of hazardous chemicals need to be evaluated and strengthened to ensure their effectiveness at each level. Process safety management is the proactive identification, evaluation and mitigation, or prevention of chemical releases that could occur as a result of failures in processes, procedures, or equipment.

The process safety management standard targets highly hazardous chemicals that have the potential to cause a catastrophic incident. The purpose of the standard as a whole is to aid employers in their efforts to prevent or mitigate episodic chemical releases that could lead to a catastrophe in the workplace and possibly in the surrounding community.

To control these types of hazards, employers need to develop the necessary expertise, experience, judgement, and initiative within their workforce to properly implement and maintain an effective process safety management program as envisioned in the Occupational Safety and Health Administration (OSHA) standard.

The OSHA standard is required by the Clean Air Act Amendments, as is the Environmental Protection Agency's Risk Management Plan, which was proposed in 1992. Employers who merge the two sets of requirements into their process safety management program will better assure full compliance with each as well as enhance their relationship with the local community.

Although OSHA believes process safety management will have a positive effect on the safety of employees and will offer other potential benefits to employers, such as increased productivity, smaller businesses which may have limited resources available to them at this time might consider alternative avenues of decreasing the risks associated with highly hazardous chemicals at their workplaces. One method that might be considered is reducing inventory of the highly hazardous chemical. This reduction in inventory will result in reducing the risk or potential for a catastrophic incident. Also, employers, including small employers, may establish more efficient inventory control by reducing, to below the established threshold, the quantities of highly hazardous chemicals onsite. This reduction can be accomplished by ordering smaller shipments and maintaining the minimum inventory necessary for efficient and safe operation. When reduced inventory is not feasible, the employer might consider dispersing inventory to several locations onsite.

Dispersing storage into locations so that a release in one location will not cause a release in another location is also a practical way to reduce the risk or potential for catastrophic incidents.

Exceptions

The PSM standard does not apply to the following:

- Retail facilities
- Oil or gas well drilling or servicing operations
- Normally unoccupied remote facilities

- Hydrocarbon fuels used solely for workplace consumption as a fuel (e.g., propane used for comfort heating, gasoline for vehicle refueling), if such fuels are not a part of a process containing another highly hazardous chemical covered by this standard
- Flammable liquids stored in atmospheric tanks or transferred, which are kept below their normal boiling point without benefit of chilling or refrigerating and are not connected to a process

Process Safety Information

Hazards of the Chemicals Used in the Process

Complete and accurate written information concerning process chemicals, process technology, and process equipment is essential to an effective process safety management program and to a process hazard analysis. The compiled information will be a necessary resource to a variety of users including the team performing the process hazard analysis as required by PSM, those developing the training programs and the operating procedures, contractors whose employees will be working with the process, those conducting the pre-startup reviews, as well as local emergency preparedness planners, and insurance and enforcement officials.

The information to be compiled about the chemicals, including process intermediates, needs to be comprehensive enough for an accurate assessment of the fire and explosion characteristics, reactivity hazards, the safety and health hazards to workers, and the corrosion and erosion effects on the process equipment and monitoring tools. Current material safety data sheet (MSDS) information can be used to help meet this requirement but must be supplemented with process chemistry information, including runaway reaction and over-pressure hazards, if applicable.

Technology of the Process

Process technology information will be a part of the process safety information package and should include employer-established criteria for maximum inventory levels for process chemicals; limits beyond which would be considered upset conditions; and a qualitative estimate of the consequences or results of deviation that could occur if operating beyond the established process limits. Employers are encouraged to use diagrams that will help users understand the process.

A block flow diagram is used to show the major process equipment and interconnecting process flow lines and flow rates, stream composition, temperatures, and pressures when necessary for clarity. The block flow diagram is a simplified diagram.

Process flow diagrams are more complex and show all main flow streams including valves to enhance the understanding of the process as well as pressures and temperatures on all feed and product lines within all major vessels and in and out of headers and heat exchangers, and points of pressure and temperature control. Also, information on construction materials, pump capacities and pressure heads, compressor horsepower, and vessel design pressures and temperatures are shown when necessary for clarity. In addition, process flow diagrams usually show major components of control loops along with key utilities.

Equipment in the Process

Piping and instrument diagrams (P&IDs) may be the more appropriate type diagrams to show some of the above details as well as display the information for the piping designer and engineering staff. The P&IDs are to be used to describe the relationships between equipment and instrumentation as well as other relevant information that will enhance clarity. Computer software programs that do P&IDs or other diagrams useful to the information package may be used to help meet this requirement.

The information pertaining to process equipment design must be documented. In other words, what codes and standards were relied on to establish good engineering practice? These codes and standards are published by such organizations as the American Society of Mechanical Engineers, the American Petroleum Institute, American National Standards Institute, National Fire Protection Association, American Society for Testing and Materials, The National Board of Boiler and Pressure Vessel Inspectors, National Association of Corrosion Engineers, American Society of Exchange Manufacturers Association, and Model Building Code groups.

For existing equipment designed and constructed many years ago in accordance with the codes and standards available at that time and no longer in general use today, the employer must document which codes and standards were used and that the design and construction along with the testing, inspection, and operation are still suitable for the intended use. Where the process technology requires a design that departs from the applicable codes and standards, the employer must document that the design and construction are suitable for the intended purpose.

Employee Involvement

Section 304 of the Clean Air Act Amendments states that employers are to consult with their employees and their representatives regarding their efforts in developing and implementing the process safety management program elements and hazard assessments. Section 304 also requires employers to train and educate their employees and to inform affected employees of the findings from incident investigations required by the process safety management program. Many employers, under their existing safety and health programs, already have established methods to keep employees and their representatives informed about relevant safety and health issues and may be able to adapt these practices and procedures to meet their obligations under PSM.

Employers who have not implemented an occupational safety and health program may wish to form a safety and health committee of employees and management representatives to help the employer meet the PSM obligations. Such a committee can be a significant ally in helping the employer implement and maintain an effective process safety management program for all employees.

Process Hazard Analysis

A process hazard analysis (PHA), or evaluation, is one of the most important elements of the process safety management program. A PHA is an organized and systematic effort to identify and analyze the significance of potential hazards associated with the processing or handling of highly hazardous chemicals. A PHA provides information that will assist employers and employees in making decisions for improving safety and reducing the consequences of unwanted or unplanned releases of hazardous chemicals.

A PHA analyzes potential causes and consequences of fires, explosions, releases of toxic or flammable chemicals, and major spills of hazardous chemicals. The PHA focuses on equipment, instrumentation, utilities, human actions (routine and nonroutine), and external factors that might affect the process.

The selection of a PHA methodology or technique will be influenced by many factors including how much is known about the process. Is it a process that has been operated for a long period of time with little or no innovation and extensive experience has been generated with its use? Or, is it a new process or one that has been changed frequently by the inclusion of innovation features? Also, the size and complexity of the process will influence the decision as to the appropriate PHA methodology to use. All PHA methodologies are subject to certain limitations. For example, the checklist methodology works well when the process is very stable and no changes are made, but it is not as effective when the process has undergone extensive change. The checklist may miss the most recent changes and consequently they would not be evaluated. Another limitation to be considered concerns the assumptions made by the team or analyst. The PHA is dependent on good judgement, and the assumptions made during the study need to be documented and understood by the team and reviewer and kept for a future PHA.

The team conducting the PHA needs to understand the methodology that is going to be used. A PHA team can vary in size from two people to a number of people with varied operational and technical backgrounds. Some team members may be part of the team for only a limited time. The team leader needs to be fully knowledgeable in the proper implementation of the PHA methodology to be used and should be impartial in the evaluation. The other full- or part-time team members need to provide the team with expertise in areas such as process technology; process design; operating procedures and practices; alarms; emergency procedures; instrumentation; maintenance procedures; both routine and nonroutine tasks, including how the tasks are authorized; procurement of parts and supplies; safety and health; and any other relevant subjects. At least one team member must be familiar with the process.

The ideal team will have an intimate knowledge of the standards, codes, specifications, and regulations applicable to the process being studied. The selected team members need to be compatible, and the team leader needs to be able to manage the team and the PHA study. The team needs to be able to work together while benefiting from the expertise of others on the team or outside the team to resolve issues and to forge a consensus on the findings of the study and recommendations.

The application of a PHA to a process may involve the use of different methodologies for various parts of the process. For example, a process involving a series of unit operations of varying sizes, complexities, and ages may use different methodologies and team members for each operation. Then the conclusions can be integrated into one final study and evaluation.

A more specific example is the use of a PHA checklist for a standard boiler or heat exchanger and the use of a Hazard and Operability PHA for the overall process. Also, for batch-type processes like custom batch operations, a generic PHA of a representative batch may be used where there are only small changes of monomer or other ingredient ratio and the chemistry is documented for the full range and ratio of batch ingredients.

Another process where the employer might consider using a generic type of PHA is a gas plant. Often these plants are simply moved from site to site, and therefore, a generic PHA may be used for these movable plants. Also, when an employer has several similar size gas plants and no sour gas is being processed at the site, a generic PHA is feasible as long as the variations of the individual sites are accounted for in the PHA.

Finally, when an employer has a large continuous process with several control rooms for different portions of the process, such as for a distillation tower and a blending operation, the employer may wish to do each segment separately and then integrate the final results.

Small businesses covered by this rule often will have processes that have less storage volume and less capacity, and may be less complicated than processes at a large facility. Therefore, OSHA would anticipate that the less complex methodologies would be used to meet the process hazard analysis criteria in the standard. These process hazard analyses can be done in less time and with fewer people being involved. A less complex process generally means that less data, P&IDs, and process information are needed to perform a process hazard analysis.

Many small businesses have processes that are not unique, such as refrigerated warehouses or cold storage lockers or water treatment facilities. Where employer associations have a number of members with such facilities, a generic PHA, evolved from a checklist or what-if questions, could be developed and effectively used by employers to reflect their particular process; this would simplify compliance for them.

When the employer has a number of processes that require a PHA, the employer must set up a priority system to determine which PHAs to conduct first. A preliminary hazard analysis may be useful in setting priorities for the processes that the employer has determined are subject to coverage by the process safety management standard. Consideration should be given first to those processes with the potential of adversely affecting the largest number of employees. This priority setting also should consider the potential severity of a chemical release, the number of potentially affected employees, the operating history of the process, such as the frequency of chemical releases, the age of the process, and any other relevant factors. Together, these factors would suggest a ranking order using either a weighting factor system or a systematic ranking method. The use of a preliminary hazard analysis will assist an employer in determining which process should be of the highest priority for hazard analysis resulting in the greatest improvement in safety at the facility occurring first.

Detailed guidance on the content and application of process hazard analysis methodologies is available from the American Institute of Chemical Engineers' Center for Chemical Process Safety, 345 E. 47th Street, New York, New York 10017, (212) 705-7319. Also, see the discussion of various methods of process hazard analysis contained in the Appendix to this publication.

Operating Procedures

Operating procedures describe tasks to be performed, data to be recorded, operating conditions to be maintained, samples to be collected, and safety and health precautions to be taken. The procedures need to be technically accurate, understandable to employees, and revised periodically to ensure that they reflect current operations. The process safety information package helps to ensure that the operating procedures and practices are consistent with the known hazards of the chemicals in the process and that the operating parameters are correct. Operating procedures should be reviewed by engineering staff and operating personnel to ensure their accuracy and that they provide practical instructions on how to actually carry out job duties safely. Also, the employer must certify annually that the operating procedures are current and accurate.

Operating procedures provide specific instructions or details on what steps are to be taken or followed in carrying out the stated procedures. The specific instructions should include the applicable safety precautions and appropriate information on safety implications. For example, the operating procedures addressing operating parameters will contain operating instructions about pressure limits, temperature ranges, flow rates, what to do when an upset condition occurs, what alarms and instruments are pertinent if an upset condition occurs, and other subjects. Another example of using operating instructions to properly implement operating procedures is in starting up or shutting down the process. In these cases, different parameters will be required from those of normal operation. These operating instructions need to clearly indicate the distinctions between startup and normal operations, such as the appropriate allowances for heating up a unit to reach the normal operating parameters. Also, the operating instructions need to describe the proper method for increasing the temperature of the unit until the normal operating temperatures are reached.

Computerized process control systems add complexity to operating instructions. These operating instructions need to describe the logic of the software as well as the relationship between the equipment and the control system; otherwise, it may not be apparent to the operator.

Operating procedures and instructions are important for training operating personnel. The operating procedures are often viewed as the standard operating practices (SOPs) for operations. Control room personnel and operating staff, in general, need to have a full understanding of operating procedures. If workers are not fluent in English, then procedures and instructions need to be prepared in a second language understood by the workers. In addition, operating procedures need to be changed when there is a change in the process. The consequences of operating procedure changes need to be fully evaluated and the information conveyed to the personnel. For example, mechanical changes to the process made by the maintenance department (such as changing a valve from steel to brass or other subtle changes) need to be evaluated to determine whether operating procedures and practices also need to be changed. All management of change actions must be coordinated and integrated with current operating procedures, and operating personnel must be alerted to the changes in procedures before the change is made. When the process is shut down to make a change, then the operating procedures must be updated before restarting the process.

Training must include instruction on how to handle upset conditions as well as what operating personnel are to do in emergencies such as pump seal failures or pipeline ruptures. Communication among operating personnel and workers within the process area performing nonroutine tasks also must be maintained. The hazards of the tasks are to be conveyed to operating personnel in accordance with established procedures and to those performing the actual tasks. When the work is completed, operating personnel should be informed to provide closure on the job.

Employee Training

All employees, including maintenance and contractor employees involved with highly hazardous chemicals, need to fully understand the safety and health hazards of the chemicals and processes they work with so they can protect themselves, their fellow employees, and the citizens of nearby communities. Training conducted in compliance with the OSHA Hazard Communication Standard (Title 29 Code of Federal Regulations (CFR) Part 1910.1200) will inform employees about the chemicals they work with and familiarize them with reading and understanding MSDSs. However, additional training in subjects such as operating procedures and safe work practices, emergency evacuation and response, safety procedures, routine and nonroutine work authorization activities, and other areas pertinent to process safety and health need to be covered by the employer's training program.

In establishing their training programs, employers must clearly identify the employees to be trained, the subjects to be covered, and the goals and objectives they wish to achieve. The learning goals or objectives should be written in clear measurable terms before the training begins. These goals and objectives need to be tailored to each of the specific training modules or segments. Employers should describe the important actions and conditions under which the employee will demonstrate competence or knowledge as well as what is acceptable performance.

Hands-on training, where employees actually apply lessons learned in simulated or real situations, will enhance learning. For example, operating personnel, who will work in a control room or at control panels, would benefit by being trained at a simulated control panel. Upset conditions of various types could be displayed on the simulator, and then the employee could go through the proper operating procedures to bring the simulator panel back to the normal operating parameters. A training environment could be created to help the trainee feel the full reality of the situation but under controlled conditions. This type of realistic training can be very effective in teaching employees correct procedures while allowing them also to see the consequences of what might happen if they do not follow established operating procedures. Other training techniques using videos or training also can be very effective for teaching other job tasks, duties, or imparting other important information. An effective training program will allow employees to fully participate in the training process and to practice their skills or knowledge.

Employers need to evaluate periodically their training programs to see if the necessary skills, knowledge, and routines are being properly understood and implemented by their trained employees. The methods for evaluating the training should be developed along with the training program goals and objectives. Training program evaluation will help employers to determine the amount of training their employees understood, and whether the desired results were obtained. If, after the evaluation, it appears that the trained employees are not at the level of knowledge and skill that was expected, the employer should revise the training program, provide retraining, or provide more frequent refresher training sessions until the deficiency is resolved. Those who conducted the training and those who received the training also should be consulted as to how best to improve the training process. If there is a language barrier, the language known to the trainees should be used to reinforce the training messages and information.

Careful consideration must be given to ensure that employees, including maintenance and contract employees, receive current and updated training. For example, if changes are made to a process, affected employees must be trained in the changes and understand the effects of the changes on their job tasks. Additionally, as already discussed, the evaluation of the employees' absorption of training will certainly determine the need for further training.

Contractors

Employers who use contractors to perform work in and around processes that involve highly hazardous chemicals have to establish a screening process so that they hire and use only contractors who accomplish the desired job tasks without compromising the safety and health of any employees at a facility. For contractors whose safety performance on the job is not known to the hiring employer, the employer must obtain information on injury and illness rates and experience and should obtain contractor references. In addition, the employer must ensure that the contractor has the appropriate job skills, knowledge, and certifications (e.g., for pressure vessel welders). Contractor work methods and experience should be evaluated. For example, does the contractor conducting demolition work swing loads over operating processes or does the contractor avoid such hazards?

Maintaining a site injury and illness log for contractors is another method employers must use to track and maintain current knowledge of activities involving contract employees working on or adjacent to processes covered by PSM. Injury and illness logs of both the employer's employees and contract employees allow the employer to have full knowledge of process injury and illness experience. This log contains information useful to those auditing process safety management compliance and those involved in incident investigations.

Contract employees must perform their work safely. Considering that contractors often perform very specialized and potentially hazardous tasks, such as confined space entry activities and nonroutine repair activities, their work must be controlled while they are on or near a process covered by PSM. A permit system or work authorization system for these activities is helpful for all affected employers. The use of a work authorization system keeps an employer informed of contract employee activities. Thus, the employer has better coordination and more management control over the work being performed in the process area. A well-run and well-maintained process, where employee safety is fully recognized, benefits all of those who work in the facility whether they are employees of the employer or the contractor.

Prestartup Safety Review

For new processes, the employer will find a PHA helpful in improving the design and construction of the process from a reliability and quality point of view. The safe operation of the new process is enhanced by making use of the PHA recommendations before final installations are completed. P&IDs should be completed, the operating procedures in place, and the operating staff trained to run the process, before startup. The initial startup procedures and normal operating procedures must be fully evaluated as part of the prestartup review to ensure a safe transfer into the normal operating mode.

For existing processes that have been shut down for turnaround or modification, the employer must ensure that any changes other than "replacement in kind" made to the process during shutdown go through the management of change procedures. P&IDs will need to be updated, as necessary, as well as operating procedures and instructions. If the changes made to the process during shutdown are significant and affect the training program, then operating personnel as well as employees engaged in routine and nonroutine work in the process area may need some refresher or additional training in light of the changes. Any incident investigation recommendations, compliance audits, or PHA recommendations need to be reviewed to see what affect they may have on the process before beginning the startup.

Mechanical Integrity of Equipment

Employers must review their maintenance programs and schedules to see if there are areas where "breakdown" maintenance is used rather than the more preferable ongoing mechanical integrity program. Equipment used to process, store, or handle highly hazardous chemicals has to be designed, constructed, installed, and maintained to minimize the risk of releases of such chemicals. This requires that a mechanical integrity program be in place to ensure the continued integrity of process equipment.

Elements of a mechanical integrity program include the identifying and categorizing equipment and instrumentation; inspections and tests and their frequency; maintenance procedures; training of maintenance personnel; criteria for acceptable test results; documentation of test and inspection results; and documentation of manufacturer recommendations for equipment and instrumentation.

Process Defenses

The first line of defense an employer has is to operate and maintain the process as designed and to contain the chemicals. This is backed up by the second line of defense which is to control the release of chemicals through venting to scrubbers or flares, or to surge or overflow tanks designed to receive such chemicals. This also would include fixed fire protection systems like sprinklers, water spray, or deluge systems, monitor guns, dikes, designed drainage systems, and other systems to control or mitigate hazardous chemicals once an unwanted release occurs.

Written Procedures

The first step of an effective mechanical integrity program is to compile and categorize a list of process equipment and instrumentation to include in the program. This list includes pressure vessels, storage tanks, process piping, relief and vent systems, fire protection system components, emergency shutdown systems and alarms, and interlocks and pumps. For the categorization of instrumentation and the listed equipment, the employer should set priorities for which pieces of equipment require closer scrutiny than others.

Inspection and Testing

The mean time to failure of various instrumentation and equipment parts would be known from the manufacturer's data or the employer's experience with the parts, which then influence inspection and testing frequency and associated procedures. Also, applicable codes and standards—such as the National Board Inspection Code, or those from the American Society for Testing and Materials, American Petroleum Institute, National Fire Protection Association, American National Standards Institute, American Society of Mechanical Engineers, and other groups—provide information to help establish an effective testing and inspection frequency, as well as appropriate methodologies.

The applicable codes and standards provide criteria for external inspections for such items as foundation and supports, anchor bolts, concrete or steel supports, guy wires, nozzles and sprinklers, pipe hangers, grounding connections, protective coatings and insulation, and external metal surfaces of piping and vessels. These codes and standards also provide information on methodologies for internal inspection and frequency formulas based on the corrosion rate of the materials of construction. Also, internal and external erosion must be considered along with corrosion effects for piping and valves. Where the corrosion rate is not known, a maximum inspection frequency is recommended (methods of developing the corrosion rate are available in the codes). Internal inspections need to cover items such as the vessel shell, bottom, and head; metallic linings; nonmetallic linings; thickness measurements for vessels and piping; inspection for erosion, corrosion, cracking, and bulges; internal equipment like trays, baffles, sensors, and screens for erosion, corrosion, or cracking and other deficiencies. Some of these inspections may be performed by state or local government inspectors under state and local statutes. However, each employer must develop procedures to ensure that tests and inspections are conducted properly and that consistency is maintained even where different employees may be involved. Appropriate training must be provided to maintenance personnel to ensure that they understand the preventive maintenance program procedures, safe practices, and the proper use and application of special equipment or unique tools that may be required.

This training is part of the overall training program called for in the standard.

Quality Assurance

A quality assurance system helps ensure the use of proper materials of construction, the proper fabrication and inspection procedures, and appropriate installation procedures that recognize field installation concerns. The quality assurance program is an essential part of the mechanical integrity program and will help maintain the primary and secondary lines of defense designed into the process to prevent unwanted chemical releases or to control or mitigate a release. "As built" drawings, together with certifications of coded vessels and other equipment and materials of construction, must be verified and retained in the quality assurance documentation.

Equipment installation jobs need to be properly inspected in the field for use of proper materials and procedures and to assure that qualified craft workers do the job. The use of appropriate gaskets, packing, bolts, valves, lubricants, and welding rods needs to be verified in the field. Also, procedures for installing safety devices need to be verified, such as the torque on the bolts on rupture disc installations, uniform torque on flange bolts, and proper installation of pump seals. If the quality of parts is a problem, it may be appropriate for the employer to conduct audits of the equipment supplier's facilities to better ensure proper purchases of required equipment suitable for intended service. Any changes in equipment that may become necessary will need to be reviewed for management of change procedures.

Nonroutine Work Authorizations

Nonroutine work conducted in process areas must be controlled by the employer in a consistent manner. The hazards identified involving the work to be accomplished must be communicated to those doing the work and to those operating personnel whose work could affect the safety of the process. A work authorization notice or permit must follow a procedure that describes the steps the maintenance supervisor, contractor representative, or other person needs to follow to obtain the necessary clearance to start the job. The work authorization procedures must reference and coordinate, as applicable, lockout/tagout procedures, line breaking procedures, confined space entry procedures, and hot work authorizations. This procedure also must provide clear steps to follow once the job is completed to provide closure for those who need to know the job is now completed and that equipment can be returned to normal.

Managing Change

To properly manage changes to process chemicals, technology, equipment, and facilities one must define what is meant by change. In the process safety management standard, change includes all modifications to equipment, procedures, raw materials, and processing conditions other than "replacement in kind." These changes must be properly managed by identifying and reviewing them prior to implementing them. For example, the operating procedures contain the operating parameters (pressure limits, temperature ranges, flow rates, etc.) and the importance of operating within these limits. While the operator must have the flexibility to maintain safe operation within the established parameters, any operation outside of these parameters requires review and approval by a written management of change procedure. Management of change also covers changes in process technology and changes to equipment and instrumentation. Changes in process technology can result from changes in production rates, raw materials, experimentation, equipment unavailability, new equipment, new product development, change in catalysts, and changes in operating conditions to improve yield or quality.

Equipment changes can be in materials of construction, equipment specifications, piping prearrangements, experimental equipment, computer program revisions, and alarms and interlocks. Employers must establish means and methods to detect both technical and mechanical changes.

Temporary changes have caused a number of catastrophes over the years, and employers must establish ways to detect both temporary and permanent changes. It is important that a time limit on temporary changes be established and monitored since otherwise, without control, these changes may tend to become permanent. Temporary changes are subject to the management of change provisions. In addition, the management of change procedures is used to ensure that the equipment and procedures are returned to their original or designed conditions at the end of the temporary change. Proper documentation and review of these changes are invaluable in ensuring that safety and health considerations are incorporated into operating procedures and processes. Employers may wish to develop a form or clearance sheet to facilitate the processing of changes through the management of change procedures. A typical change form may include a description and the purpose of the change, the technical basis for the change, safety and health considerations, documentation of changes for the operating procedures, maintenance procedures, inspection and testing, P&IDs, electrical classification, training and communications, prestartup inspection, duration (if a temporary change), approvals, and authorization. Where the impact of the change is minor and well understood, a checklist reviewed by an authorized person, with proper communication to others who are affected, may suffice.

For a more complex or significant design change, however, a hazard evaluation procedure with approvals by operations, maintenance, and safety departments may be appropriate. Changes in documents such as P&IDs, raw materials, operating procedures, mechanical integrity programs, and electrical classifications should be noted so that these revisions can be made permanent when the drawings and procedure manuals are updated. Copies of process changes must be kept in an accessible location to ensure that design changes are available to operating personnel as well as to PHA team members when a PHA is being prepared or being updated.

Incident Investigation

Incident investigation is the process of identifying the underlying causes of incidents and implementing steps to prevent similar events from occurring. The intent of an incident investigation is for employers to learn from past experiences and thus avoid repeating past mistakes. The incidents OSHA expects employers to recognize and to investigate are the types of events that resulted in or could reasonably have resulted in a catastrophic release. These events are sometimes referred to as "near misses," meaning that a serious consequence did not occur but could have.

Employers must develop in-house capability to investigate incidents that occur in their facilities. A team should be assembled by the employer and trained in the techniques of investigation, including how to conduct interviews of witnesses, assemble needed documentation, and write reports. A multidisciplinary team is better able to gather the facts of the event and to analyze them and develop plausible scenarios as to what happened and why. Team members should be selected on the basis of their training, knowledge, and ability to contribute to a team effort to fully investigate the incident.

Employees in the process area where the incident occurred should be consulted, interviewed, or made members of the team. Their knowledge of the events represents a significant set of facts about the incident that occurred. The report, its findings, and recommendations should be shared with those who can benefit from the information. The cooperation of employees is essential to an effective incident investigation. The focus of the investigation should be to obtain facts, and not to place blame. The team and the investigative process should clearly deal with all involved individuals in a fair, open, and consistent manner.

Emergency Preparedness

Each employer must address what actions employees are to take when there is an unwanted release of highly hazardous chemicals. Emergency preparedness is the employer's third line of defense that will be relied on along with the second line of defense, which is to control the release of chemicals. Control releases and emergency preparedness will take place when the first line of defense to operate and maintain the process and contain the chemicals fails to stop the release. In preparing for an emergency chemical release, employers will need to decide the following:

- Whether they want employees to handle and stop small or minor incidental releases;
- Whether they wish to mobilize the available resources at the plant and have them brought to bear on a more significant release;
- Whether employers want their employees to evacuate the danger area and promptly escape to a preplanned safe zone area, and then allow the local community emergency response organizations to handle the release; or
- Whether the employer wants to use some combination of these actions.

Employers will need to select how many different emergency preparedness or third lines of defense they plan to have, develop the necessary emergency plans and procedures, appropriately train employees in their emergency duties and responsibilities, and then implement these lines of defense.

Employers, at a minimum, must have an emergency action plan that will facilitate the prompt evacuation of employees when there is an unwanted release of a highly hazardous chemical. This means that the employer's plan will be activated by an alarm system to alert employees when to evacuate, and that employees who are physically impaired will have the necessary support and assistance to get them to a safe zone. The intent of these requirements is to alert and move employees quickly to a safe zone. The use of process control centers or buildings as safe areas is discouraged. Recent catastrophes indicate that lives are lost in these structures because of their location and because they are not necessarily designed to withstand over-pressures from shock waves resulting from explosions in the process area.

When there are unwanted incidental releases of highly hazardous chemicals in the process area, the employer must inform employees of the actions/procedures to take. If the employer wants employees to evacuate the area, then the emergency action plan will be activated. For outdoor processes, where wind direction is important for selecting the safe route to a refuge area, the employers should place a wind direction indicator, such as a wind sock or pennant, at the highest point visible throughout the process area. Employees can move upwind of the release to gain safe access to the refuge area by knowing the wind direction.

If the employer wants specific employees in the release area to control or stop the minor emergency or incidental release, these actions must be planned in advance and procedures developed and implemented. Handling incidental releases for minor emergencies in the process area must include preplanning, providing appropriate equipment for the hazards, and conducting training for those employees who will perform the emergency work before they respond to handle an actual release. The employer's training program, including the Hazard Communication Standard training, is to address, identify, and meet the training needs for employees who are expected to handle incidental or minor releases.

Preplanning for more serious releases is an important element in the employer's line of defense. When a serious release of a highly hazardous chemical occurs, the employer, through preplanning, will have determined in advance what actions employees are to take. The evacuation of the immediate release area and other areas, as necessary, would be accomplished under the emergency action plan. If the employer wishes to use plant personnel—such as a fire brigade, spill control team, a hazardous materials team—or employees to render aid to those in the immediate release area and to control or mitigate the incident, refer to OSHA's Hazardous Waste Operations and Emergency Response (HAZWOPER) standard (Title 29 CFR Part 1910.120). If outside assistance is necessary, such as through mutual aid agreements between employers and local government emergency response organizations, these emergency responders are also covered by HAZWOPER. The safety and health protection required for emergency responders is the responsibility of their employers and of the on-scene incident commander.

Responders may be working under very hazardous conditions; therefore, the objective is to have them competently led by an on-scene incident commander and the commander's staff, properly equipped to do their assigned work safely, and fully trained to carry out their duties safely before they respond to an emergency. Drills, training exercises, or simulations with the local community emergency response planners and responder organizations is one means to obtain better preparedness. This close cooperation and coordination between plant and local community emergency preparedness managers also will aid the employer in complying with the Environmental Protection Agency's Risk Management Plan criteria. (EPA is required by the Clean Air Act Amendments of 1990 to develop regulations that will require companies to make available to the public information on the way the companies manage the risks of the chemicals they handle. These regulations will be developed in 1992. The OSHA PSM standard, which meets similar Clean Air Amendment requirements and the forthcoming EPA rules, will apply only to specified chemicals in listed quantities. OSHA and EPA's lists will not necessarily be identical.)

An effective way for medium to large facilities to enhance coordination and communication during emergencies within the plant and with local community organizations is by establishing and equipping an emergency control center. The emergency control center would be located in a safe zone so that it could be occupied throughout the duration of an emergency. The center should serve as the major communications link between the on-scene incident commander and plant or corporate management as well as with local community officials. The communications equipment in the emergency control center should include a network to receive and transmit information by telephone, radio, or other means. It is important to have a backup communication network in case of power failure or if one communication means fails. The center also should be equipped with the plant layout; community maps; utility drawings, including water for fire extinguishing; emergency lighting; appropriate reference materials such as a government agency notification list, company personnel phone list, SARA Title III reports and material safety data sheets, emergency plans and procedures manual; a listing with the location of emergency response equipment and mutual aid information; and access to meteorological or weather condition data and any dispersion modeling data.

Compliance Audits

An audit is a technique used to gather sufficient facts and information, including statistical information, to verify compliance with standards. Employers must select a trained individual or assemble a trained team to audit the process safety management system and program. A small process or plant may need only one knowledgeable person to conduct an audit. The audit includes an evaluation of the design and effectiveness of the process safety management system and a field inspection of the safety and health conditions and practices to verify that the employer's systems are effectively implemented. The audit should be conducted or led by a person knowledgeable in audit techniques who is impartial towards the facility or area being audited. The essential elements of an audit program include planning, staffing, conducting the audit, evaluating hazards and deficiencies and taking corrective action, performing a follow-up, and documenting actions taken.

Planning

Planning is essential to the success of the auditing process. During planning, auditors should select a sufficient number of processes to give a high degree of confidence that the audit reflects the overall level of compliance with the standard. Each employer must establish the format, staffing, scheduling, and verification methods before conducting the audit. The format should be designed to provide the lead auditor with a procedure or checklist that details the requirements of each section of the standard. The names of the audit team members should be listed as part of the format as well. The checklist, if properly designed, could serve as the verification sheet that provides the auditor with the necessary information to expedite the review of the program and ensure that all requirements of the standard are met. This verification sheet format could also identify those elements that will require an evaluation or a response to correct deficiencies. This sheet also could be used for developing the follow-up and documentation requirements.

Staffing

The selection of effective audit team members is critical to the success of the program. Team members should be chosen for their experience, knowledge, and training and should be familiar with the processes and auditing techniques, practices and procedures.

The size of the team will vary depending on the size and complexity of the process under consideration. For a large, complex, highly instrumented plant, it may be desirable to have team members with expertise in process engineering and design; process chemistry; instrumentation and computer controls; electrical hazards and classifications; safety and health disciplines; maintenance; emergency preparedness; warehousing or shipping; and process safety auditing. The team may use part-time members to provide the expertise required and to compare what is actually done or followed with the written PSM program.

Conducting the Audit

An effective audit includes a review of the relevant documentation and process safety information, inspection of the physical facilities, and interviews with all levels of plant personnel. Utilizing the audit procedure and checklist developed in the preplanning stage, the audit team can systematically analyze compliance with the provisions of the standard and any other corporate policies that are relevant. For example, the audit team will review all aspects of the training program as part of the overall audit. The team will review the written training program for adequacy of content, frequency of training, effectiveness of training in terms of its goals and objectives as well as to how it fits into meeting the standard's requirements. Through interviews, the team can determine employees' knowledge and awareness of the safety procedures, duties, rules, and emergency response assignments. During the inspection, the team can observe actual practices such as safety and health policies, procedures, and work authorization practices. This approach enables the team to identify deficiencies and determine where corrective actions or improvements are necessary.

Evaluation and Corrective Action

The audit team, through its systematic analysis, should document areas that require corrective action as well as where the process safety management system is effective. This provides a record of the audit procedures and findings and serves as a baseline of operation data for future audits. It will assist in determining changes or trends in future audits.

Corrective action is one of the most important parts of the audit and includes identifying deficiencies, and planning, following up, and documenting the corrections. The corrective action process normally begins with a management review of the audit findings. The purpose of this review is to determine what actions are appropriate, and to establish priorities, timetables, resource allocations and requirements, and responsibilities. In some cases, corrective action may involve a simple change in procedures or a minor maintenance effort to remedy the problem. Management of change procedures need to be used, as appropriate, even for a seemingly minor change. Many of the deficiencies can be acted on promptly, while some may require engineering studies or more detailed review of actual procedures and practices. There may be instances where no action is necessary; this is a valid response to an audit finding. All actions taken, including an explanation when no action is taken on a finding, need to be documented.

The employer must assure that each deficiency identified is addressed, the corrective action to be taken is noted, and the responsible audit person or team is properly documented. To control the corrective action process, the employer should consider the use of a tracking system. This tracking system might include periodic status reports shared with affected levels of management, specific reports such as completion of an engineering study, and a final implementation report to provide closure for audit findings that have been through management of change, if appropriate, and then shared with affected employees and management. This type of tracking system provides the employer with the status of the corrective action. It also provides the documentation required to verify that appropriate corrective actions were taken on deficiencies identified in the audit.

Conclusion

OSHA believes the preceding discussion of PSM should help small employers to comply more easily with the new requirements the standard imposes. The end result can only be a safer, more healthful workplace for all employees—a goal we all share.

APPENDIX: METHODS OF PROCESS HAZARD ANALYSIS

On July 17, 1990, OSHA issued a proposed rule for the management of hazards associated with processes using highly hazardous chemicals. This rule, called the Process Safety Management Standard, was finalized on February 24, 1992. In an appendix to the proposed rule, OSHA discussed several methods of process hazard analysis. That discussion may be helpful for those doing job hazard analyses. Refer to Chapter 4 for these methods.

RELATED PUBLICATIONS

OSHA-2056 "All About OSHA."
OSHA-3084 "Chemical Hazard Communication."
OSHA-3047 "Consultation Services for the Employer."
OSHA-3088 "How to Prepare for Workplace Emergencies."
OSHA-2098 "OSHA Inspections."
OSHA-3021 "OSHA: Employee Workplace Rights."
OSHA-3077 "Personal Protective Equipment."
OSHA-3132 "Process Safety Management."
OSHA-3079 "Respiratory Protection."

Hazard Communication Standard, Title 29, *Code of Federal Regulations* (CFR) Part 1910.1200.

Process Safety Management of Highly Hazardous Chemicals Standard, Title 29, *Code of Federal Regulations* (CFR) Part 1910.119 FR 57, P. 6356. This contains the actual text of the PSM rule.

(A single free copy of the above materials can be obtained from OSHA Publications Office, Room N3101, Washington, D.C. 20210, (202) 219-9667).

OSHA-3104 "Hazard Communication: A Compliance Kit" (A reference guide to step-by-step requirements for compliance with the OSHA standard).

OSHA-3071 "Job Hazard Analysis."

(OSHA 3104 and OSHA 3071 are available from the Superintendent of Documents, U.S. Government Printing Office, Washington, D.C. 20402, (202) 783-3238. OSHA 3104—GPO Order No. 929-022-000009; $18—domestic, $22.50—foreign. OSHA 3071—GPO Order No. 029-016-00142-5, $1.00.)

Appendix D

Site Audit Subjects

There are two main reasons that we have included this information with this book. We believe this information will assist the reader in ensuring that the safety plan is being implemented effectively. And also, it will give the reader an idea of what an OSHA compliance officer will be looking for should a compliance inspection take place at their site. Be advised that the information is geared toward hazardous waste incinerator sites. However, this information has applicability toward the larger hazardous waste site whether an incinerator is on site or not.

In Chapter 7 we talked about implementation of a safety plan. We believe that this information will be useful in determining the effectiveness of the HASP. Keep in mind that review of the HASP and its effectiveness is part of the HAZWOPER standard. Besides being part of the standard, we believe it is just good business. The likelihood of your safety program running well and for the long term will increase as those responsible continue to audit and make appropriate adjustments or "tuning."

For larger facilities, we believe that a site-specific audit form should be developed by a group of qualified professionals who are familiar with the site. The well-written safety plan can provide a sort of "skeleton" for the audit sections. For each area of the site, a list of audit questions (and we prefer yes or no answers) should be developed. There are a variety of ways that this audit can be used. Some sections or all of the audit might be used by an in-house safety committee while performing periodic inspections. Or possibly the corporate safety or quality departments might work through this audit on a yearly basis. The users and time frame for use will vary according to the hazards, size of the site, and talent of available people.

As one can imagine, the audit process can be people intensive and expensive. And besides the audit process being expensive, one might find that findings from the audit indicate that certain actions are recommended. At times, to implement these actions can be complicated, controversial, expensive, and time consuming.

The authors believe that site audits should be site specific. The audit protocol and specific areas to be inspected should be designed with one specific site in mind. Although the site-specific audit protocol is highly recommended, it is rarely adopted. Time and money constraints will, many times, not allow for site-specific audit protocols to be developed. What typically will occur is that a "standard" audit form will be used.

Using a standard audit form is certainly acceptable and can prove to be an outstanding tool, especially when getting a handle on a fire extinguisher program, life-safety issues, the lighting of exit signs, availability of first-aid kits, clear aisles, and a variety of issues that are basic in nature. The typical standard audit form will likely have complete sections that are marked "not applicable." Unless a unique audit form is created, there is usually no way around sections that will be marked "not applicable."

For small sites that are active for a relatively short duration, it is impractical to even attempt a site-specific audit form. But for manufacturing facilities, or sites that are large or of long duration, the development of a site-specific audit form should be considered. We suggest that you consider the following information when developing site-specific audits, which is excerpted from OSHA and was designed specifically for hazardous waste sites that contained an incinerator. Sites that contain an incinerator are usually considered long term as far as duration, and one would likely be dealing with highly hazardous substances. This type of site should be considered for a site-specific rather than standard audit form.

What we have included is only a small part of a much longer document available in its entirety at the address shown or on the Web. We provide the complete table of contents so that you get a good idea of the subjects covered and the amount of detail taken to cover them. After the table of contents, we have taken several sections applying to working with hazardous substances. We believe that OSHA has chosen these items to ensure a safe and healthful workplace. COSHOs will use this format when performing an OSHA compliance audit on incinerator sites.

Occupational Safety and Health Administration
Directorate of Compliance Programs
200 Constitution Avenue, N.W.
Washington, D.C. 20210

Table of Contents

INTRODUCTION
GENERAL OVERVIEW ON INCINERATOR TECHNOLOGY
  AND ASSOCIATED HAZARDS
CONDUCT OF THE INSPECTION

SAFETY AND HEALTH AUDIT QUESTIONS
Safety and Health Program
Site Control
Training
Medical Monitoring
Engineering Controls, Work Practices, and Personal Protective Equipment for Employee Protection
Monitoring
Decontamination
Emergency Response
Heat Stress Program
Hotwork Fire Prevention and Protection
Lockout/Tagout
Confined Space Program
Incinerator Process Safety
Review of the Site's Safety and Health Plan (SAHP)

The site's SAHP is the most important document for the inspection, because it describes all elements of the site's program; this document will thus drive much of the inspection process and serve as a reference point for the walkaround phase of the inspection. The following section discusses the steps to be followed by the inspection team reviewing the SAHP.

The inspection team should obtain a copy of the site's most recent SAHP. Because conditions on the hazardous waste site change so rapidly, it is important that the SAHP be current; OSHA's experience shows that it is not uncommon for the site SAHP to be out of date. At a minimum, the SAHP should address the following elements:

- Names of key personnel responsible for site safety;
- Safety and health risk analysis for each site task and operation;
- Site control measures;
- Employee training assignments;
- Medical surveillance requirements;
- Personal protective equipment for each of the site tasks and operations;
- Frequency and types of air monitoring, personnel monitoring, environmental sampling techniques, and instrumentation, along with methods for maintenance and calibration of equipment;
- Evaluation of site for presence of chemicals in the amounts requiring process safety management procedures;
- Confined space entry procedures;
- Spill containment program;
- Decontamination procedures; and
- Emergency response plan.

1910.120 (d): SITE CONTROL
1910.120 (b)(4)(ii)(B), WRITTEN SITE CONTROL PROGRAM
1910.120 (d), SITE CONTROL

III. Verification of Program Elements

A. Records Review

1. Does the SAHP contain site control procedures that have been developed during the planning stages of a hazardous waste clean-up operation and modified as new information becomes available? [OSHA Reference .120(b)(4)(ii)(F), .120(d)(2)]
2. Does the site control program include, as a minimum, the following (where these requirements are covered elsewhere they need not be repeated); [OSHA Reference .120(d)(3)]
   - a site map;
   - site work zones;
   - use of a "buddy system";
   - site communication including alerting means for emergencies; [OSHA Reference .120(d)(3)]
   - standard operating procedures or safe work practices; and
   - identification of the nearest medical assistance?

B. On-site Conditions

1. Are work zones including Exclusion Zone (EZ), Contamination Reduction Zone (CRZ), and Support Zones adequately demarcated and is restricted access enforced? [OSHA Reference .120(d)(3)]
   Are the observed locations of Zones and the methods of demarcation as described in the SAHP? [OSHA Reference .120(b)(4)(ii)(F)]
2. Is the "buddy system" rigorously adhered to in areas identified by the SAHP? (The buddy system is defined as a system of organizing employees into work groups such that each employee in the group is designated to be observed by at least one other employee in the group). [OSHA Reference .120(d)(3)]
3. Is the nearest medical assistance readily identified? [OSHA Reference .120(d)(3)]
   Is the information, including telephone numbers, addresses, and location of medical assistance conspicuously posted in the Control Zone?

C. Interviews

1. Are work zones including Exclusion Zone (EZ), Contamination Reduction Zone (CRZ), and Support Zones adequately demarcated and is restricted access enforced? [OSHA Reference .120(d)(3)]

2. Is the "buddy system" rigorously adhered to in areas identified by the SAHP? [OSHA Reference .120(d)(3)]
    Are work groups formally designated or are employees simply instructed to "watch out for each other"?
3. Do you know where the nearest medical assistance is and how to request it? [OSHA Reference .120(d)(3)]
4. Are employees aware of the existence and location of SOPs for safely performing job tasks? [OSHA Reference .120(d)(3)]

1910.120(e): TRAINING
1910.120(b)(1)(ii)(D), WRITTEN SAFETY AND HEALTH TRAINING PROGRAM
1910.120(b)(4)(ii)(B), TRAINING ELEMENT OF SAHP
1910.120(e), TRAINING

III. Verification of Program Elements

A. Records Review

1. Has the employer developed a written safety and health training program? [OSHA Reference .120(b)(1)(ii)(D)]
2. Has the written program been incorporated as part of the SAHP? [OSHA Reference .120(b)(4)(ii)(B)]
3. Do the elements of the training program include at least the following: [OSHA Reference .120(e)(2)]
    - Names of personnel and alternates responsible for site safety and health; [OSHA Reference (i)]
    - Safety, health, and other hazards on the site; [OSHA Reference (ii)]
    - Use of personal protective equipment; [OSHA Reference (iii)]
    - Work practices used to minimize hazards; [OSHA Reference (iv)]
    - Safe use of engineering controls and equipment on the site; [OSHA Reference (v)]
    - Medical surveillance requirements, including recognition of symptoms and signs that might indicate overexposure to hazards; and [OSHA Reference (vi)]
    - The contents of the SAHP? [OSHA Reference (vii)]
4. Do the SAHP and personnel records demonstrate that:
    - Employees receive training before they are permitted to engage in hazardous waste operations; [OSHA Reference .120(e)(1)]
    - General site workers receive a minimum of 40 hours of off-site instruction and three days of supervised on-site training; [OSHA Reference .120(b)(4)(ii)(B), (e)(3)(i), and (e)(3), (iv)]
    - Workers assigned specific limited tasks receive at least 24 hours of off-site instruction and one day of supervised on-site training; [OSHA Reference .120(b)(4)(ii)(B) and (e)(3)(ii)]

- Workers who work in well-characterized areas, who are not exposed above permissible limits, and where there is no possibility of an emergency receive at least 24 hours of off-site instruction and one day of supervised on-site training; [OSHA Reference .120(b)(4)(ii)(B) and (e)(3)(iii)]
- On-site management and supervisors receive an additional 8 hours of specialized training at the time of job assignment; [OSHA Reference .120(b)(4)(ii)(B) and (e)(4)]
- Trainers are qualified to instruct employees and have satisfactorily completed a training program for teachers or have necessary academic credentials or instructional experience; [OSHA Reference .120(b)(4)(ii)(B) and (e)(5)]
- Employees and supervisors have been issued written certificates by their instructor and trained supervisor; [OSHA Reference .120(b)(4)(ii)(B) and (e)(7)]
- Employees engaged in responding to emergency situations have been trained in how to respond to such situations; [OSHA Reference .120(b)(4)(ii)(B) and (e)(7)]
- Employees and supervisors receive at least 8 hours of refresher training each year; and [OSHA Reference .120(b)(4)(ii)(B) and (e)(8)]
- Employees and supervisors who have not had initial training can show by documentation or certification evidence of equivalent training or work experience? [OSHA Reference .120(b)(4)(ii)(B) and (e)(9)]

B. On-site Conditions

1. Do employees appear to be aware of safety, health, and other hazards present on site? [OSHA Reference .120(e)(2)(ii)]
2. Do employees appear to have been trained in the use of personal protective equipment? [OSHA Reference .120(e)(2)(iii)]
3. Are employees implementing work practices that can minimize the risks from hazards? [OSHA Reference .120(e)(2)(iv)]
4. Do employees appear to be trained in the safe use of engineering controls and equipment? [OSHA Reference .120(e)(2)(v)]

C. Interviews

1. Have employees received training before engaging in hazardous waste operations? [OSHA Reference .120(e)(1)]
2. Have employees received training in the following:
    - Safety, health, and other hazards present on site; [OSHA Reference .120(e)(2)(ii)]

- Use of personal protective equipment; [OSHA Reference .120(e)(2)(iii)]
- Work practices that can minimize the risk of hazards; and [OSHA Reference .120(e)(2)(iv)]
- Safe use of engineering controls and equipment? [OSHA Reference .120(e)(2)(v)]
3. Are employees familiar with medical surveillance requirements and recognition of signs and symptoms that indicate overexposure to hazards (including signs and symptoms of heat stress)? [OSHA Reference .120(e)(2)(vi)]
4. Are employees familiar with the contents of the site SAHP? [OSHA Reference .120(e)(2)(vii)]

1910.120 (f): MEDICAL MONITORING
1910.120 (b)(4)(ii)(D), WRITTEN MEDICAL SURVEILLANCE PROGRAM
1910.120 (f), MEDICAL SURVEILLANCE

III. Verification of Program Elements

A. Records Review

1. Do the SAHP and site records indicate that a medical surveillance program has been established for the following type of employees: [OSHA Reference .120(b)(4)(ii)(D)]
    - All employees who are or may be exposed to hazardous substances or health hazards at or above PELs or other published exposure levels without regard to the use of respirators for 30 or more days per year; [OSHA Reference .120(f)(1), (f)(2)(i)]
    - All employees who wear a respirator for 30 days or more a year or as required by 1910.134; [OSHA Reference .120(f)(1), (f)(2)(ii)]
    - All employees who are injured, become ill or develop signs or symptoms due to possible overexposure involving hazardous substances or health hazards from an emergency response or hazardous waste operation; and [OSHA Reference .120(f)(1), (f)(2)(iii)]
    - All members of a HAZMAT team? [OSHA Reference .120(f)(1), (f)(2)(iv)]
2. Do the SAHP and site records indicate that the medical surveillance program requires medical exams and consultations on the following schedules: [OSHA Reference .120(b)(4)(ii)(D)]
    For employees included in categories A, B, or D, above: [OSHA Reference .120(f)(3)]
    - Prior to assignment; [OSHA Reference .120(i)(a)]
    - At least once every 12 months or biennially (with physicians's approval); [OSHA Reference .120(i)(b)]

- At termination of employment or reassignment to an area where an employee would not be covered, unless the employee has had an exam within the last 6 months; [OSHA Reference .120(i)(c)]
- When an employee has developed signs or symptoms indicating possible overexposure; [OSHA Reference .120(i)(d)]
- When employee has been injured or exposed above a PEL in an emergency situation; and [OSHA Reference .120(i)(d)]
- At more frequent times on medical advice? [OSHA Reference .120(i)(e)]

For employees in category C above:
- As soon as possible following the emergency incident or development of signs or symptoms; and [OSHA Reference .120(ii)(a)]
- At additional times, if the examining physician determines that follow-up examinations or consultations are medically necessary? [OSHA Reference .120(ii)(b)]

3. Do the SAHP and site records indicate that the medical surveillance program provides for exams to contain the following: [OSHA Reference .120(b)(4)(ii)(D)]
    - Medical and work history with special emphasis on symptoms related to the handling of hazardous substances and health hazards, and; [OSHA Reference .120(f)(4)(i)]
    - Fitness for duty including the ability to wear any required PPE under conditions that may be expected at the work site (i.e., temperature extremes)? [OSHA Reference .120(f)(4)(I)]
4. Do the SAHP and site records require that the content of medical examinations or consultations made available to the employees shall be determined by the attending physician under the guidelines of the "Four Agency Hazardous Waste Document"? [OSHA Reference .120(b)(4)(ii)(D), (f)(4)(ii)]
5. Do the SAHP and site records indicate that the medical exams and procedures are: [OSHA Reference .120(b)(4)(ii)(d)]
    - Performed by or under the supervision of a licensed physician; and [OSHA Reference .120(f)(5)]
    - Provided to the employee; [OSHA Reference .120(f)(5)] without cost to the employee; without loss of pay; and at a reasonable time and place?
6. Do the SAHP and medical records indicate that the employer is providing the following to the examining physician: [OSHA Reference .120(b)(4)(ii)(D)]
    - A copy of this standard and its appendices; [OSHA Reference .120(f)(6)]
    - A description of the employee's duties as they relate to the employee's exposure; [OSHA Reference .120(f)(6)(i)]
    - The employee's exposure levels or anticipated exposure levels; [OSHA Reference .120(f)(6)(ii)]

- A description of any personal protective equipment used or to be used; [OSHA Reference .120(f)(6)(iii)]
- Information from previous medical examinations of the employee which is not readily available to the examining physician; and [OSHA Reference .120(f)(6)(iv)]
- Information required by 1910.134? [OSHA Reference .120(f)(6)(v)]

7. Do the SAHP and site records indicate that the employer obtains and furnishes to the employee a copy of a written opinion from the attending physician that contains the following: [OSHA Reference .120(b)(4)(ii)(D)]
   - The physician's opinion as to whether the employee has any detected medical conditions which would place the employee at increased risk of material impairment of the employee's health from work in hazardous waste operations or a emergency response, or from respirator use; [OSHA Reference .120(f)(7)(i)(A)]
   - The physician's recommended limitations upon the employee's assigned work; [OSHA Reference .120(f)(7)(i)(B)]
   - The results of the medical examination and tests if requested by the employee; [OSHA Reference .120(f)(7)(i)(C)]
   - A statement that the employee has been informed by the physician of the results of the medical examination and any medical conditions which require further examination or treatment; and [OSHA Reference .120(f)(7)(i)(D)]
   - The written opinion shall not reveal specific findings or diagnoses unrelated to occupational exposures? [OSHA Reference .120(f)(7)(ii)]

8. Do the SAHP and site records indicate that medical surveillance records are retained as specified by 1910.20, and do the retained records contain the following: [OSHA Reference .120(b)(4)(ii)(D)]
   - Name and Social Security number of the employee; [OSHA Reference .120(f)(8)(ii)(A)]
   - Physician's written opinions, recommended limitations, and results of examinations and tests; [OSHA Reference .120(ii)(B)]
   - Any employee medical complaints related to exposure to hazardous substances; and [OSHA Reference .120(ii)(C)]
   - A copy of the information provided to the examining physician by the employer? [OSHA Reference .120(ii)(D)]

B. *On-site Conditions*

Not Applicable

C. *Interviews*

1. Is the employee covered by the medical surveillance requirements? If so:

- Has the employee been examined at the correct frequency; [OSHA Reference .120(f)(3)]
- Has the employee seen the physician's written opinion; [OSHA Reference .120(f)(7)]
- Has there been a situation where the employee should have been examined and has not; and [OSHA Reference .120(f)(2)]
- Has the employee had the exams at no cost or loss of pay and at a reasonable time and place? [OSHA Reference .120(f)(5)]

If not, should the employee be covered by the surveillance? [OSHA Reference .120(f)(2)]

1910.120: (b): SAFETY AND HEALTH PROGRAM
1910.120(b), SAFETY AND HEALTH PROGRAM

III. Verification of Program Elements

A. Records Review

1. Does the employer have an up-to-date written safety and health plan (SAHP)? [OSHA Reference .120(b)(1)(i)]
2. Does the SAHP contain each of the following elements:
    - An organizational structure; [OSHA Reference .120(b)(1)(ii)(A)]
    - A comprehensive workplan; and [OSHA Reference .120(b)(1)(ii)(B)]
    - A site-specific safety and health plan that includes the employer's standard operating procedures (SOPs) for safety and health? [OSHA Reference .120(b)(1)(ii)(C), (F)]
3. Does the SAHP include a means of informing subcontractors of site emergency procedures and health and safety hazards present on site? [OSHA Reference .120(b)(1)(iv)]
4. Is the written SAHP readily available to: [OSHA Reference .120(b)(1)(v)]
    - Employees and their representatives;
    - Subcontractors and their employees; and
    - OSHA personnel or other federal, state, or local agencies with regulatory authority?
5. Does the organizational part of the SAHP establish a specific chain of command and specify responsibilities of supervisors and employees? [OSHA Reference .120(b)(2)(i)]

    Are health and safety personnel (including alternates) identified in the SAHP? [OSHA Reference .120(b)(2)(ii)]

    Are the names and titles of individuals identified in the organizational structure current?

    Does the organizational structure include at least:

- A general supervisor who directs all hazardous waste operations; [OSHA Reference .120(b)(2)(i)(A)]
- A site safety and health supervisor who has authority to develop and implement the plan; [OSHA Reference .120(b)(2)(i)(B)]
- General functions and responsibilities of all other personnel needed for hazardous waste activities; and [OSHA Reference .120.(b)(2)(i)(D)]
- The lines of authority, responsibility, and communication?

6. Does the comprehensive workplan address the tasks and objectives of site operations? [OSHA Reference .120(b)(3)]

   Does the comprehensive workplan address anticipated cleanup activities and normal operating procedures? [OSHA Reference .120(b)(3)(i)]

7. Does the site-specific safety and health plan include at least the following elements:
   - A safety and health risk analysis for each task and operation performed on site; [OSHA Reference .120(b)(4)(ii)(A)]
   - Employee training program; [OSHA Reference .120(ii)(B)]
   - A written personal protective equipment (PPE) program; [OSHA Reference .120(ii)(C)]
   - A written medical surveillance program, including identification of all employees entered in the program, a description of medical examinations and tests routinely administered, identification of the physician in charge of the program, and a description of record-keeping procedures; [OSHA Reference .120(ii)(D)]
   - A written monitoring program that describes the frequency and types of air monitoring to be conducted, instrumentation used, and methods for maintenance and calibration of equipment; [OSHA Reference .120(ii)(E)]
   - A description of site control measures; [OSHA Reference .120(ii)(F)]
   - Written decontamination procedures; [OSHA Reference .120(ii)(G)]
   - A written emergency response plan; [OSHA Reference .120(ii)(H)]
   - Confined space entry procedures; and [OSHA Reference .120 (ii)(I)]
   - A spill containment program? [OSHA Reference .120(ii)(J)]

8. Do the job-specific safety and health analyses contain specific information on the nature of safety and health hazards associated with each job performed on site, and do they provide specific instructions to employees for avoiding the hazards? [OSHA Reference .120(b)(4)(ii)(A)]

   Does the SAHP describe the principal chemical contaminants, affected media, anticipated or measured concentrations, potential routes of exposure, and health effects associated with exposure to the contaminants?

Does the SAHP identify the appropriate level of PPE for each site task and operation?

B. On-site Conditions

1. Are the SAHP and job-specific safety and health analyses readily available to employees in the Control Zone and other accessible areas? [OSHA Reference .120(b)(1)(v)]

C. Interviews

1. Are copies of the SAHP and job-specific safety and health analyses readily available? [OSHA Reference .120(b)(1)(v)]

1910.120 (g): ENGINEERING CONTROLS, WORK PRACTICES, AND PERSONAL PROTECTIVE EQUIPMENT FOR EMPLOYEE PROTECTION
1910.120(b)(4)(ii)(C), WRITTEN PERSONAL PROTECTIVE EQUIPMENT PROGRAM
1910.120(g), ENGINEERING CONTROLS, WORK PRACTICES, AND PPE

III. Verification of Program Elements

A. Records Review

1. Is the employer using means other than employee rotation to comply with permissible exposure limits (PELs) or ionizing radiation dose limits, except where no other feasible way exists? [OSHA Reference .120(g)(1)(iii)]
2. Does the SAHP contain a written personal protective equipment (PPE) program which addresses the following: [OSHA Reference .120(b)(4); (ii)(C) and (g)(5)]
   - PPE selection based on site hazards; [OSHA Reference .120 (g)(5)]
   - PPE use and limitations of the equipment; [OSHA Reference .120(i)]
   - Work mission duration; [OSHA Reference .120(iii)]
   - PPE maintenance and storage; [OSHA Reference .120(iv)]
   - PPE decontamination and disposal; [OSHA Reference .120(v)]
   - PPE training and proper fitting; [OSHA Reference .120(vi)]
   - PPE donning and doffing procedures; [OSHA Reference .120(vii)]
   - PPE inspection procedures prior to, during, and after use; [OSHA Reference .120(viii)]
   - Evaluation of the effectiveness of the PPE program; and [OSHA Reference .120(ix)]

- Appropriate medical considerations, such as limitations during temperature extremes and potential for heat stress? [OSHA Reference .120(x)]
3. Personal protective equipment selection: [OSHA Reference .120(g)(3)]

    Is PPE selected and used to protect employees from the hazards and potential hazards they are likely to encounter as identified during the site characterization and analysis (including physical hazards such as heat stress, ionizing radiation, and noise)?

    Is PPE selected and used to meet the requirements of 29 CFR Part 1910, Subpart I (eye and face protection, respiratory protection, occupational head protection, occupational foot protection, and electrical protection devices)?

    Is PPE selected based on an evaluation of performance characteristics of the PPE relative to requirements and limitations of the site, task-specific conditions and duration, and hazards and potential hazards identified?

    Has the employer conducted any objective monitoring (i.e., of contamination of the skin or work clothes) to evaluate the effectiveness of PPE selected? [OSHA Reference .120(g)(3)]

    Is positive pressure self-contained breathing apparatus (SCBA) or positive pressure air-line respirator and escape air supply used when chemical exposure will create a substantial possibility of immediate death, immediate serious illness or injury, or impair the ability to escape?

    Are totally encapsulating chemical protective suits (Level A) used in conditions where skin absorption of a hazardous substance may result in a substantial possibility of immediate death, immediate serious illness or injury, or impair the ability to escape?

    Is the level of PPE increased when additional information indicates that increased protection is necessary to reduce employee exposure below PELs and published exposure levels?

    Does the site safety and health officer have the authority to upgrade the required level of PPE when site conditions warrant?

    Does the site safety and health officer have the authority to downgrade the required level of PPE, when it is safe to do so, to reduce the potential for heat stress?
4. Are totally encapsulating chemical protective suits: [OSHA Reference .120(g)(4)]
    - Selected to protect employees for hazards identified during site characterization and analysis;
    - Capable of maintaining positive air pressure; and [OSHA Reference .120(g)(4)]
    - Capable of preventing inward test gas leakage of more than 0.5%?

5. If applicable, has the employer implemented a hearing conservation program that includes noise monitoring, use of hearing protection devices, and audiograms? [OSHA Reference 1910.95(c)]

B. On-site Conditions

1. Has the employer implemented the use of engineering controls (e.g., pressurized cabs or control booths, remotely operated material handling equipment) and work practices (e.g., removing all nonessential personnel during drum opening, wetting down dusty operations, working upwind of possible inhalation hazards) to reduce and maintain employee exposure to or below permissible exposure limits to the extent feasible? [OSHA Reference .120(g)(1)(i)]
2. Does the employer comply with 29 CFR, Subpart G (OSHA standards for ventilation, noise, and ionizing and nonionizing radiation)? [OSHA Reference .120(g)(1)(iv)]
3. Has the employer implemented the use of engineering controls, work practices, and personal protective equipment to reduce and maintain employee exposure to or below published exposure levels for hazardous substances and health hazards not regulated by 29 CFR Part 1910, Subparts G and Z (e.g., heat stress, lifting hazards)? [OSHA Reference .120(g)(2)]
4. Personal protective equipment selection: [OSHA Reference .120(g)(3)]

    Does PPE appear to have been selected and used to protect employees from the hazards and potential hazards they are likely to encounter?

    Does the potential for heat stress appear to have been considered in the selection of PPE?

    Is positive pressure self-contained breathing apparatus (SCBA) or positive pressure air-line respirator and escape air supply used when chemical exposure will create a substantial possibility of immediate death, immediate serious illness or injury, or impair the ability to escape?

    Are totally encapsulating chemical protective suits (Level A) used in conditions where skin absorption of a hazardous substance may result in a substantial possibility of immediate death, immediate serious illness or injury, or impair the ability to escape?

    Is PPE selected and used to meet the requirements of 29 CFR Part 1910, Subpart I (eye and face protection, respiratory protection, occupational head protection, occupational foot protection, and electrical protection devices)?

    Is PPE selected in accordance with the written program contained in the SAHP? [OSHA Reference .120(b)(4)(ii)(C)]

C. *Interviews*

1. Does the employer implement the use of engineering controls (e.g., pressurized cabs or control booths, remotely operated material handling equipment) and work practices (e.g., removing all nonessential personnel during drum opening, wetting down dusty operations, working upwind of possible inhalation hazards) to reduce and maintain employee exposure to or below permissible exposure limits to the extent feasible? [OSHA Reference .120(g)(1)(i)]
2. Does the employer use means other than employee rotation to comply with permissible exposure limits (PELs) or radiation dose limits except where no other feasible way exists? [OSHA Reference .120(g)(1)(iii)]
3. Are employees familiar with the types of PPE included in Levels A, B, C, and D ensembles, as appropriate for the site? [OSHA Reference .120(g)(3)]
4. Are employees familiar with procedures for inspecting, maintaining, cleaning, and disposing of PPE? [OSHA Reference .120(g)(5)]
5. Have employees ever encountered situations that indicate that their PPE is not protecting them from exposure (i.e., respirator failure or leakage of moisture through protective clothing)? [OSHA Reference .120(g)(3)]

1910.120:(h): MONITORING
1910.120(b)(4)(ii)(E), WRITTEN MONITORING PROGRAM
1910.120(h), MONITORING
1910.95, OCCUPATIONAL NOISE EXPOSURE
1910.96, IONIZING RADIATION

III. Verification of Program Elements

A. *Records Review*

1. Does the SAHP contain a program or procedures to monitor employee exposures to all hazardous substances known or suspected on site? [OSHA Reference .120(b)(4)(ii)(E)]
2. Does the monitoring program or procedures contain the following: [OSHA Reference .120(b)(4)(ii)(E)]
    Frequency and types of:
   - air monitoring;
   - personnel monitoring;
   - environmental sampling (for heat stress, noise, radiation), including: [OSHA Reference .120(b)(4)(ii)(E)]
   - sampling techniques;
   - instrumentation;

- types; and
- methods of calibration and maintenance?

3. Does the SAHP contain monitoring requirements and procedures to be conducted after prior monitoring when: [OSHA Reference .120(h)(3)]
    - The possibility of an IDLH condition or flammable atmosphere has developed; or
    - There is an indication that exposures may have risen over PELs under such conditions as:
    - When work begins on a different portion of the site; [OSHA Reference .120(h)(3)(i)]
    - When the contaminants other than those previously identified are being handled; [OSHA Reference .120(h)(3)(ii)]
    - When a different type of operation is initiated; or [OSHA Reference .120(h)(3)(iii)]
    - When employees are handling leaking drums or containers or working in areas with obvious liquid contamination? [OSHA Reference .120(h)(3)(iv)]
4. Does the SAHP prescribe personal monitoring programs to meet the specific personal monitoring requirements for materials present on site that are listed in 1910.1001–1048? [OSHA Reference 1910.1001 to .1048, as applicable]
5. Does the SAHP require that the employees who are likely to have the highest exposure to hazardous substances and health hazards are monitored by using personal sampling frequently enough to adequately characterize employee exposures? [OSHA Reference .120(h)(4)]
6. When the exposures of employees likely to have the highest exposure are over the PELs or other published exposure levels, does the SAHP require that monitoring shall continue to determine the exposures of all employees likely to be above those limits? [OSHA Reference .120(h)(4)]
7. Are the sampling and monitoring methods used appropriate for the substances identified? [OSHA Reference .120(h)(1)]
8. Is the sampling frequency appropriate for the work task and the substances identified? [OSHA Reference .120(h)(1)]
9. Is a qualified laboratory used to analyze exposure samples? [OSHA Reference .120(h)(1)]
10. Are sampling and monitoring results returned in a reasonable time frame to prevent harm to employees if the results are above PELs or published exposure levels? [OSHA Reference .120(h)(1)]
11. Are sampling and monitoring results identified as to personal or area locations? [OSHA Reference.120(h)(1)]
12. Are sampling and monitoring results used to determine the appropriate level of employee protection needed on site? [OSHA Reference .120(h)(1)]

13. Are the maintenance and calibration procedures for the sampling and monitoring instrumentation adequate to assure accurate results? [OSHA Reference .120(h)(1)]
14. Are real-time monitoring instrument results correctly correlated to sampling results? [OSHA Reference .120(h)(1)]
15. Are the correct "indicator substances" used to characterize employee exposures to the hazardous substances to which they are exposed? [OSHA Reference .120(h)(1)]
    Is monitoring being routinely conducted for the indicator substances identified in the SAHP? [OSHA Reference .120(b)(4)(ii)(E)]
16. Does the SAHP contain specific procedures to respond to overexposures detected from monitoring? [OSHA Reference .120(h)(1)]
17. Are there up-to-date maintenance and calibration logs for all sampling and monitoring instruments? [OSHA Reference .120(h)(1)]

B. *On-site Conditions*

1. Are sampling and monitoring being performed correctly regarding?: [OSHA Reference .120(h)(1)
    - Location of samples or readings
    - Instrument operation
    - Analysis of results or readings
    - Recording of results or readings
2. Is instrument calibration performed appropriately? [OSHA Reference .120(h)(1)]

C. *Interviews*

1. Does the person performing the sampling and monitoring have sufficient training to: [OSHA Reference.120(h)(1)]
    - Assure accurate results?
    - Assure proper response to overexposure results?
2. Are employees notified of their monitoring results? [OSHA Reference.120(i), Informational programs]
3. Do employees understand the significance or meaning of the monitoring results? [OSHA Reference .120(i), Informational programs]

1910.120 (k): DECONTAMINATION
1910.120(b)(4)(ii)(G), WRITTEN DECONTAMINATION PROGRAM
1910.120 (k), DECONTAMINATION

III. Verification of Program Elements

A. *Records Review*

1. Does the SAHP contain procedures for all phases of decontamination (decon) including:

- Method of communicating procedures to employees before allowing them to enter the site; [OSHA Reference .120(k)(2)(i)]
- SOPs that address methods for minimizing employee contact with hazardous substances or contaminated equipment; [OSHA Reference .120(k)(2)(ii)]
- Decontamination of employees leaving a contaminated area; [OSHA Reference .120(k)(2)(iii)]
- Decontamination or disposal of clothing or equipment leaving a contaminated area; [OSHA Reference .120(k)(2)(iii)]
- Decontamination or disposal of equipment and solvents used for decon; [OSHA Reference .120(k)(4)]
- Monitoring of decon procedures by site safety and health supervisor to determine effectiveness; [OSHA Reference .120(k)(2)(iv)]
- Steps to be taken when deficiencies are found; [OSHA Reference .120(k)(2)(iv)]
- Location of decon areas to minimize exposure to uncontaminated employees or equipment; [OSHA Reference .120(k)(3)]
- Decon, cleaning, laundering, maintenance, or replacement of protective clothing and equipment; [OSHA Reference .120(k)(5)(i)]
- Steps to be taken when non-impermeable clothing is splashed by contaminated materials; [OSHA Reference .120(k)(5)(i)]
- Unauthorized removal of equipment or protective clothing from change rooms; [OSHA Reference .120(k)(6)]
- Informing commercial laundries of potentially harmful effects of contaminated PPE, if applicable; and [OSHA Reference .120(k)(7)]
- Showers and change rooms? [OSHA Reference .120(k)(8)]

B. On-site Conditions

1. Do employees follow procedures that minimize contact with hazardous substances or contaminated equipment? [OSHA Reference .120(k)(2)(i)]
2. Are employees appropriately decontaminated before leaving contaminated area? [OSHA Reference.120(k)(2)(iii)]
    Is all contaminated clothing and equipment leaving a contaminated area disposed of or decontaminated appropriately?
3. Are all equipment and solvents used for decon decontaminated or disposed of properly? [OSHA Reference .120(k)(4)]
4. Is decon performed in areas that minimize the exposure of uncontaminated employees or equipment (i.e., from runoff or overspray)? [OSHA Reference .120(k)(3)]
5. Are the following requirements of 1910.141(d)(3) met: [OSHA Reference .141(d)(3)]

- One shower provided for each ten employees;
- Hot and cold water feeding on discharge line;
- Individual clean towels; and
- Body soap and cleansing agents?
6. Are change rooms provided as per 1910.141(e)? [OSHA Reference .141(e)]
    - Separate storage for street clothes and protective clothing; and
    - For cleanup operations of six months or more duration, are two separate change areas separated by a shower area provided?
7. Are showers and change rooms located in areas where exposures are below the PELs and published exposure levels? [OSHA Reference .141(d)(3)(v) and .141(e)]

C. Interviews

1. Is the decontamination procedure communicated to employees and implemented before any employee or equipment enters areas on site where potential for exposure to hazardous substances exists? [OSHA Reference .120(k)(2)(i)]
2. Do SOPs exist which describe procedures to minimize employee contact with hazardous substances or contaminated equipment? [OSHA Reference .120(k)(2)(ii)]
3. Are decon procedures monitored to determine their effectiveness? Are you aware of any steps taken to correct deficiencies? [OSHA Reference .120(k)(2)(iv)]
4. When non-impermeable clothing becomes wetted with hazardous substances, is it immediately removed and do you shower? Is the clothing disposed of or decontaminated before it is removed from the work zone? [OSHA Reference .120(k)(5)(ii)]
5. Do unauthorized employees ever remove protective clothing or equipment from change rooms? [OSHA Reference .120(k)(6)]

1910.120(l): EMERGENCY RESPONSE
1910.120(b)(4)(iii)(h), SAFETY AND HEALTH PROGRAM
1910.120(e)(7), TRAINING, EMERGENCY RESPONSE
1910.120(l), EMERGENCY RESPONSE
1910.165, EMPLOYEE ALARM SYSTEMS

III. Verification of Program Elements

A. Records Review

1. Does the written emergency response plan in the SAHP consider all anticipated emergencies? [OSHA Reference .120(b)(4)(iii)(H), (l)(1)(i)]

2. Does the SAHP contain an emergency response plan that includes all of the following required elements: [OSHA Reference .120(b)(4)(iii)(H), (e)(7), (l)(2)]
   - Pre-emergency planning;
   - Personnel roles, lines of authority, and communication;
   - Emergency recognition and prevention;
   - Safe distances and places of refuge; [OSHA Reference .120(b)(4)(iii)(H), (e)(7), (l)(2)]
   - Site security and control;
   - Evacuation routes and procedures;
   - Decontamination procedures not covered elsewhere in the plan;
   - Emergency medical treatment and first aid;
   - Emergency alerting and response procedures;
   - PPE and emergency equipment;
   - Site topography, layout, and prevailing weather conditions;
   - Procedures for reporting incidents to local, state, and Federal government agencies; and [OSHA Reference .120(l)(3)(i)(A)]
   - Critique of response drills with follow-up? [OSHA Reference .120(l)(3)(i)(B)]
3. Is the written emergency response plan contained in a separate section of the SAHP? [OSHA Reference .120(l)(3)(ii)]
4. Does the SAHP provide for regular rehearsal of emergency response procedures as part of the overall training for emergency response? [OSHA Reference .120(l)(3)(iv), .120(l)(3)(v)]
5. Is the emergency response plan reviewed periodically and regularly updated? [OSHA Reference .120(l)(3)(vi)]
6. Does the SAHP describe an emergency response alarm system? If so:
   - Is other than voice communication used as a means of sounding the alarm (Note: voice communication is permitted on sites with 10 or fewer facilities); [OSHA Reference .165(b)(5)]
   - Are spare alarm devices and components that are subject to wear available for prompt restoration of the system; [OSHA Reference .165(c)(2)]
   - Are back-up means of alarm, such as employee runners or telephone, provided when the system is out of service; and [OSHA Reference .165(d)(3)]
   - Does the alarm system provide positive notification whenever a deficiency exists in the system? [OSHA Reference .165(d)(4)]

B. On-site Conditions

1. Does contact with local emergency responders (i.e., fire and rescue, local hospital) and local, state, and Federal agencies indicate that the emergency response plan is compatible and integrated with the dis-

aster, fire, and/or emergency response plans of those organizations? [OSHA Reference .120(l)(3)(iii)]

Have local emergency responders been provided and have readily available a copy of the site's emergency response plan?

Do local emergency responders have procedures for rescuing and/or treating personnel who are potentially contaminated? [OSHA Reference .120(l)(3)(iii)]

Have local emergency responders been provided with information on the nature of hazardous substances present at the site and the potential hazards associated with exposure to those substances?

Have local emergency responders participated in rehearsals or drills of emergency situations?
2. When telephones serve as a means of reporting emergencies, are emergency telephone numbers posted near telephones, employee notice boards, or other conspicuous locations? [OSHA Reference .165(b)(4)]
3. Are there suitable facilities for emergency flushing of the eyes and body located near areas where hazardous materials such as acids or caustics are stored (in particular, near the waste water treatment plant)? [OSHA Reference .151(c)]

C. Interviews

1. Are employees who are designated to respond to emergencies trained in how to respond to such expected emergencies? [OSHA Reference .120(e)(7)]
2. Are employees aware of personnel roles, lines of authority, and communication procedures? [OSHA Reference .120(l)(2)(ii)]
3. Do employees know all evacuation routes and procedures and the locations of places of refuge? [OSHA Reference .120(l)(2)(iv) and (vi)]
4. Do employees know decontamination procedures that are to be followed in the event of an emergency? [OSHA Reference .120(l)(2)(vii)]
5. Have employees participated in rehearsals of emergency situations? [OSHA Reference .120(l)(3)(iv)]
6. Do employees know the meaning of emergency alarm signals, as described in the SAHP? [OSHA Reference .120(l)(3)(vi)]
7. Do employees know the locations of emergency telephone numbers? [OSHA Reference .165(b)(4)]
8. Can the emergency alarm be perceived above ambient noise or light levels? [OSHA Reference .165(b)(2)]
9. Is the emergency alarm distinctive and recognizable as a signal to evacuate the work area? [OSHA Reference .165(b)(3)]
10. Is the alarm tested at least annually for reliability and adequacy? [OSHA Reference .165(d)(4)]

Heat Stress Program

A. Records Review

1. Is there a written heat stress prevention program as part of the SAHP or safety and health SOPs?
   Does the program include the following elements:
   - Environmental monitoring for heat stress conditions;
   - Provision for selecting appropriate PPE to minimize the risk of heat stress;
   - Biologic monitoring for signs of heat stress (including pulse rate, oral temperature, and/or blood pressure measurements);
   - Implementation of work/rest schedules based on the results of environmental monitoring;
   - Provision for cool rest areas, including shelters within the exclusion zone;
   - A liquid replacement program; and
   - An acclimatization program?
2. Has the employer implemented a heat stress training program?
3. Does the employer regularly monitor heat conditions (i.e., dry bulb or adjusted dry bulb temperatures) to determine the risk of heat stress and to establish appropriate work/rest regimens? (Note: Wet bulb globe temperature is not the most appropriate measure of environmental heat conditions when employees are wearing vapor impermeable protective clothing.)
4. Does the employer monitor the temperature, blood pressure, and pulse rate of employees exposed to heat stress environments?
   Do environmental heat measurements trigger implementation of physiologic monitoring?
   Are physiologic measurements taken during rest breaks and used to modify work/rest schedules?
5. Has the employer established procedures for providing medical attention or rapid cool-down for employees subject to heat stress?
6. Do the employer's OSHA 200 Log and OSHA 101 forms indicate any heat stress problems?

B. On-site Conditions

1. Does the employer have the necessary equipment to monitor employees' temperatures, blood pressures, and pulse rates?
2. Does the employer have a mechanism for informing employees of the work/rest regimen or modification of that regimen based on changed conditions?
3. Do workers in the exclusion zone have ready access to drinking water supplies, shaded rest areas, and/or air conditioned or fan-cooled areas?

4. Does the personal protective equipment selected for employees in the exclusion zone take account of the need to reduce heat stress while also providing protection from chemical and other hazards at the site?
5. Are work operations scheduled to avoid physically demanding work during periods of extreme heat?
6. Does the employer provide tools and equipment that reduce the physical demands on workers who are required to work in extreme heat conditions while wearing personal protective equipment?

C. Interviews

1. Are employees familiar with the signs of heat stress? Have they received training in how to recognize and avoid heat stress?
2. Is a work/rest regimen regularly followed when work must be performed under conditions of heat stress? Are employees regularly notified of the work/rest regimen and any changes in that regimen?
3. Are cool-down areas and drinking water supplies readily available to employees working in the exclusion zone?
4. Have employees ever informed site management that they have experienced signs and symptoms of heat stress?

1910.252(a): HOTWORK FIRE PREVENTION AND PROTECTION
1910.252(a), WELDING AND BURNING FIRE PREVENTION AND PROTECTION

III. Verification of Program Elements

A. Records Review

1. Does the SAHP establish procedures for cutting and welding in other than specifically designated areas, based on the fire potentials of plant facilities? [OSHA Reference .252(a)(2)(xiii)(A)]
2. Does the SAHP designate an individual responsible for authorizing cutting and welding operations in areas not specifically designed for such processes? [OSHA Reference .252(a)(2)(xiii)(B)]
3. Does the SAHP provide for the individual responsible for authorizing cutting and welding operations to issue written permits granting such authorization? [OSHA Reference .252(a)(2)(iv)]
4. Do permits specify precautions to be followed during cutting or welding in areas not specifically designed for such processes? [OSHA Reference .252(a)(2)(iv)]
5. Does the SAHP provide that cutters or welders and their supervisors are suitably trained in the safe operation of their equipment and the safe use of the process? [OSHA Reference .252(a)(2)(viii)(C)]

6. Does the SAHP provide for advising all subcontractors about all flammable materials or hazardous conditions of which they may not be aware? [OSHA Reference .252(a)(2)(xiii)(D)]
7. Does the SAHP provide that fire watchers be trained in the use of fire extinguishing equipment? [OSHA Reference .252(a)(2)(iii)(B)]

B. On-site Conditions

1. Does the individual responsible for authorizing hot work operations inspect the area before cutting or welding is performed? [OSHA Reference .252(a)(2)(iv)]
   Does this individual designate precautions to be followed in the form of a written permit?
   Is the hot work permit conspicuously posted in the area in which work is being performed?
2. In those instances when objects to be cut or welded are moveable and the facility has an area specifically designated for cutting and welding, are objects taken to the designated area before hot work operations are performed? [OSHA Reference .252(a)(2)(xiii)(A)]
3. If objects to be welded or cut cannot readily be moved, are all movable fire hazards in the vicinity taken to a safe place?
   If fire hazards cannot be moved to a safe place, are guarding devices used to confine heat, sparks, and slag and to protect the immovable fire hazards? [OSHA Reference .252(a)(1)(i) and (ii)]
4. In those instances when objects to be welded or cut cannot be moved and all fire hazards cannot be removed, are special precautions taken to protect combustibles from ignition sources? [OSHA Reference .252(a)(2)]
5. Are precautions taken to ensure that floor openings or cracks in the flooring are closed? [OSHA Reference .252(a)(2)(i)]
   If this is not possible, are precautions taken to ensure that any readily combustible materials on the floor below the hot work operation are not exposed to sparks?
   Are similar precautions also taken with regard to cracks or holes in walls, open doorways, and open or broken windows?
6. Is suitable fire extinguishing equipment ready for instant use? (Such equipment may consist of pails of water, buckets of sand, a hose, or portable extinguishers, depending on the nature and quantity of the combustible material exposed.) [OSHA Reference .252(a)(2)(ii)]
7. Are fire watchers on duty during, and for at least a half hour after completion of, hot work operations performed in the vicinity of combustible materials or in locations where conditions could result in other than a minor fire? [OSHA Reference .252(a)(2)(iii)]
   Do fire watchers have fire extinguishing equipment readily available?

8. Are combustibles relocated to at least 35 feet from the work site? [OSHA Reference .252(a)(2)(vii)]
    Where relocation is impracticable, are combustibles protected with flameproof covers or otherwise shielded?
9. Does the supervisor take steps to ensure that combustibles are moved or properly shielded during hot work operations? [OSHA Reference .252(a)(2)(xiv)(C)]
    Does the supervisor ensure that hot work operations are scheduled so that plant activities that might expose combustibles to ignition are not begun during hot work operations?
    Does the supervisor secure authorization for hot work operations from the designated management representative?

C. *Interviews*

1. Are employees familiar with the hot work requirements contained in the site SAHP or SOPs? [OSHA Reference .120(b)(1)(v)]
    Do employees know the identities of supervisors or others authorized to issue hot work permits?

1910.147: LOCKOUT/TAGOUT
1910.147 CONTROL OF HAZARDOUS ENERGY (LOCKOUT/TAGOUT)

III. Verification of Program Elements

A. *Records Review*

1. Does the SAHP contain a lockout/tagout program that includes energy control procedures and employee training practices? [OSHA Reference .147(c)(1)]
2. Do the lockout/tagout procedures clearly outline the scope, purpose, authorization, rules, and techniques to be utilized for the control of hazardous energy and the means of enforcing compliance? [OSHA Reference .147(c)(4)(ii)]
    Do the procedures include:
    - A specific statement of the intended use of the procedures;
    - Specific procedural steps for shutting down, isolating, blocking and securing machines or equipment to control hazardous energy;
    - Specific procedural steps for the placement, removal, and transfer of lockout or tagout devices and the responsibility for them; and [OSHA Reference .147(c)(4)(ii)]
    - Specific requirements for testing a machine or equipment to determine and verify the effectiveness of lockout devices, tagout devices, and other energy control measures?

3. Are there any energy isolating devices for which the employer's program utilizes tagout instead of lockout procedures? [OSHA Reference .147(c)(3)(ii)]

    If so, has the employer demonstrated that the tagout system achieves a level of safety equivalent to that obtained by a lockout program?
4. Has the employer certified that periodic inspection of the energy control procedures is conducted at least annually to ensure that the procedures and the requirements of the lockout/tagout standard are being followed? [OSHA Reference .147(c)(6)]
5. Do inspections include a review, between the inspector and each authorized employee, of that employee's responsibilities under the lockout/tagout program? [OSHA Reference .147(c)(6)(i)(B)]
6. Do inspection records identify the machine or equipment inspected, the date of the inspection, the employees included in the inspection, and the person performing the inspection? [OSHA Reference .147(c)(6)(ii)]
7. Does the lockout/tagout program include certification of employee training, including each employee's name and dates of training? [OSHA Reference .147(c)(7)(i) and (iv)]
8. Does each employee who is authorized to lock or tag out machines or equipment receive training in the following areas: [OSHA Reference .147(c)(7)(i)(A)]
    - Recognition of applicable hazardous energy sources;
    - The type and magnitude of the energy available in the workplace; and
    - The methods and means necessary for energy isolation and control?
9. Are all other employees who may be affected by lockout/tagout procedures instructed in the purpose and use of those procedures? [OSHA Reference .147(c)(7)(i)(B)]
10. Are other employees whose work operations are, or may be, in an area where lockout/tagout procedures may be used instructed about the procedures and the prohibitions relating to attempting to restart or reenergize machines or equipment that is locked or tagged out? [OSHA Reference .147(c)(7)(i)(C)]
11. When tagout systems are used instead of lockout, are employees trained in the limitations of tags (i.e., that they are warning devices and do not act as a lock)? [OSHA Reference .147(c)(7)(ii)]
12. Do employees receive retraining whenever there is a change in their job assignment, machines or processes, or lockout/tagout procedures? [OSHA Reference .147(c)(7)(iii)]
13. Does the employer's lockout/tagout program include specific procedures and training for those cases when the employee who applied the lockout or tagout device is not available to remove it and the

device is removed under the direction of the employer? [OSHA Reference .147(e)(3)]

B. On-site Conditions

1. Are locks, tags, and other protective materials and hardware for securing machines (e.g., chains, wedges, key blocks, adapter pins, self-locking fasteners) provided by the employer? [OSHA Reference .147(c)(5)]
2. Are lockout/tagout devices in good condition, clearly identified, standardized, and durable? [OSHA Reference .147(c)(5)(ii)]
3. Do lockout and tagout devices indicate the identity of the employee applying the devices? [OSHA .147(c)(5)(ii)(D)]
4. Do tagout devices warn against hazardous conditions if the machine or equipment is restarted or energized? Do they contain a legend such as "Do not start," etc? [OSHA Reference .147(c)(5)(iii)]
5. Is lockout/tagout implemented only by authorized employees? [OSHA Reference .147(c)(8)]
6. Are all affected employees given prior notification of the application and removal of lockout and tagout devices? [OSHA Reference .147(c)(9)]
7. Are machines shut down in an orderly fashion before energy isolating devices are locked out or tagged so as to avoid any hazards to employees as a result of equipment deenergization? [OSHA Reference.147(d)(2)]
8. Are lockout and tagout devices properly applied to energy isolating devices? [OSHA .147(d)(4)]
    Are lockout devices affixed so as to hold the energy isolating device in a "safe" or "off" position?
    Are tagout devices affixed to the energy isolating device or, when this is not possible, as close as safely possible?
9. When tagout devices are used on energy isolating devices that cannot be locked out, are additional safety measures used to ensure full employee protection? (Additional safety measures include the removal of an isolating circuit element, blocking of a controlling switch, opening of an extra disconnecting device, or the removal of a valve handle to reduce the likelihood of inadvertent energization.) [OSHA Reference .147(c)(3)(ii)]
10. Following the application of lockout or tagout devices, is all stored or residual energy relieved, disconnected, restrained, or otherwise rendered safe? [OSHA Reference .147(d)(5)]
11. Does the authorized employee verify that isolation and deenergization of the machine or equipment has been accomplished before servicing or maintenance of the machine or equipment is begun? [OSHA Reference .147(d)(6)]

12. Before lockout and tagout devices are removed from machines or equipment, is the work area inspected to ensure that: [OSHA Reference .147(e)(3)]
    - All non-essential items have been removed;
    - all machine or equipment components are operationally intact; and
    - all employees have been safely positioned or removed?
13. Are all affected employees notified before lockout or tagout devices are removed? [OSHA Reference .147(e)(ii)]
14. Are all lockout and tagout devices removed only by the employee who applied the device? [OSHA Reference .147(e)(3)]

    If the employee is not available to remove lockout and tagout devices, are the devices removed under the direction of the employer pursuant to specific procedures contained in the employer's lockout/tagout program?

    In such cases, does the employer ensure that the authorized employee has this knowledge before resuming work at that facility?
15. Are proper safety procedures followed in cases where lockout or tagout devices must be temporarily removed to test or position the machine or equipment during servicing or maintenance? [OSHA Reference .147(f)]
16. In cases where outside servicing personnel (subcontractors) are involved in servicing or maintenance activities, do the on-site employer and the outside employer inform each other of their respective lockout or tagout procedures? [OSHA Reference .147(f)(2)]

    Does the on-site employer ensure that on-site personnel understand and comply with the outside employer's lockout/tagout procedures?
17. When servicing or maintenance is performed by a crew or other group, are group lockout or tagout devices used? [OSHA Reference .147(f)(3)]

    Is one authorized employee designated as primarily responsible for a set number of employees?

    Does the lockout/tagout program include procedures for that employee to ascertain the exposure status of individual group members with regard to the lockout or tagout of machines and equipment?

    Does each authorized employee affix a personal lockout or tagout device to the group device?
18. During shift or personnel changes, are specific procedures followed to ensure the continuity of lockout or tagout protection? [OSHA Reference .147(f)(4)]

C. *Interviews*

1. Have employees received training in lockout and tagout procedures? [OSHA Reference .147(c)(7)]

Have employees ever been retrained in lockout/tagout procedures because of a change in job assignment, machines, or processes?
2. Are there times when equipment is tagged but not locked out during servicing or maintenance? [OSHA Reference .147(c)(3)(ii)]
3. Are all affected employees notified when lockout/tagout is applied for servicing or maintenance and when locks and tags are removed and machines are restarted? [OSHA Reference .147(c)(9)]
4. When servicing or maintenance is performed by a crew or other group of workers, are group lockout and tagout devices used? [OSHA Reference .147(f)(3)]

Appendix E

Commonly Used Acronyms

ACGIH	American Conference of Governmental Industrial Hygienists
AIHA	American Industrial Hygiene Association
ALARA	As Low as Reasonably Achievable
ANSI	American National Standards Institute
BMP	Best Management Practices
CDC	Centers for Disease Control
CERCLA	Comprehensive Environmental Response, Compensation and Liability Act (also known as Superfund)
CFR	Code of Federal Regulations
CPR	Cardiopulmonary Resuscitation
CRC	Contamination Reduction Corridor
CRZ	Contamination Reduction Zone
CSP	Certified Safety Professional
DHHS	Department of Health and Human Services
D&D	Decontamination and Dismantlement
DOE	Department of Energy
DOT	Department of Transportation
EAP	Emergency Action Plan
EKG	Electrocardiogram
EPA	Environmental Protection Agency
ER	Environmental Restoration
ERMC	Environmental Remediation Management Contractor
ERP	Emergency Response Plan
HASP	Health and Safety Plan
HAZMAT	Hazardous Material
HAZWOPER	Hazardous Waste Operations and Emergency Response
HEPA	High Efficiency Particulate Air
HSM	Health and Safety Manager
IDLH	Immediately Dangerous to Life or Health
JHA	Job Hazard Analyses
JSA	Job Safety Analysis
LEL/LFL	Lower Explosive Limit/Lower Flammable Limit
M&O	Contractor Management and Operations Contractor

Commonly Used Acronyms

MSDS	Material Safety Data Sheets
MSHA	Mine Safety and Health Administration
NIEHS	National Institute of Environmental Health Sciences
NIOSH	National Institute for Occupational Safety and Health
NRC	Nuclear Regulatory Commission
OSH	Occupational Safety and Health
OSHA	Occupational Safety and Health Administration
OTA	Office of Technology Assessment
OU	Operable Unit
PC	Protective Clothing
PEL	Permissible Exposure Limits
PHA	Process Hazard Analysis
PM	Project Manager
PPE	Personal Protective Equipment
PRP	Potentially Responsible Parties
RCRA	Resource Conservation and Recovery Act
REL	Recommended Exposure Limits
R&D	Research and development
SARA	Superfund Amendments and Reauthorization Act
SCBA	Self-Contained Breathing Apparatus
SHO	Safety and Health Officer
SM	Site Manager
SOP	Standard Operating Procedure
SOSG	Standard Operating Safety Guide
SSHO	Site Safety and Health Officer
SSO	Site Safety Officer
TLV	Threshold Limit Value
TLV-STEL	Threshold Limit Value-Short-Term Exposure Limit
TLV-TWA	Threshold Limit Value-Time-Weighted Average
TSD	Treatment, Storage, and Disposal
UEL/UFL	Upper Explosive Limit/Upper Flammable Limit
USCG	United States Coast Guard

Index

Absorption, 78
Action levels, 60
Administrative controls, 80
Air monitoring, 60–61
Airborne dust, 62–63
Application, 17
Approval process, 38, 74, 75

Biological hazards, 78
Bloodborne pathogens, 36
Brownfields, 5
Buddy system, 81

Change order, 220
Chemical handling procedures, 62–63
Chemical hazard control, 80
Chemical hazards, 78
Clean air lock, 161
Clean room, 161–162
Client review, 41
Colorimetric detector tubes, 60–61
Contamination reduction zone/corridor (CRZ/C), 64–65, 157, 159
Contractor agreements, 24, 213–227
Contractors/Subcontractors, 7, 29, 30, 37, 213–226

Decontamination, 10–11, 34
Decontamination procedures, 81–82, 149–163
Direct reading instruments, 59, 60
Dirty air lock, 161
Dirty room, 161
Disinfection, 156
Disqualification (Contractor), 224
Disposable PPE, 82
Dose, 61
Dust suppression, 67

Emergency action plan (EAP), 171–172
Emergency equipment, 174–175
Emergency medical care, 34
Emergency phone numbers, 34
Emergency preparedness, 11, 164–176
Emergency response, 11, 25, 165–168
Emergency response plan (ERP), 172
Emergency response training, 101
Emergency transportation, 34
Emergency treatment, 87–88
Enforcement, 90
End of Service Life Indicator (ESLI), 145
Engineering controls, 39, 40, 80
Equipment decontamination, 160
Evacuation routes, 34

Exclusion zone (EZ), 63, 82, 214
Exposure assessment, 9, 38, 65
Exposure monitoring, 60, 80

Failure mode and effect analysis (FMEA), 51
Fault tree analysis (FTA), 52
Field test kits, 61
First aid, 34
Fit test, 146
Foot/hand protection, 147

Hazard assessment, 107
Hazard-based approach, 6, 38
Hazard characterization, 9, 38, 65
Hazard communication, 80
Hazard control, 8
Hazard exposure, 19, 59
Hazard identification, 47
Hazard and Operability Study (HAZOP), 51
Head protection, 147
Health and safety manager (HSM), 36
Health and safety program, 54
Health and safety plan (HASP), 10, 54–95
HEPA, 141–142, 153, 160
Host organization, 214
Hotline, 64

Incident command system (ICS), 173
Incipient, 166
Ingestion, 78
Inhalation, 78
Injection, 78
Instructor/trainer qualification, 101
Insurance certificate, 219

Job Hazard Analysis (JHA), 42–53, 58, 79, 91, 159
    discussion method, 45
    observation method, 45
Job Safety Analysis (JSA), 58

Leachate, 63
Lead, 69, 80
Lessons learned, 39, 40, 66, 102, 123, 124, 147
Levels of protection, 63

Medical clearance, 34
Medical surveillance, 11, 83–87, 145
Memoranda of agreement (MOA), 165
Memoranda of understanding (MOU), 165
Mixed waste, 1
Monitoring instruments, 35
Morale, 7

Near hits, 41, 44, 48
Near misses, 41
Noise dosimeter, 61
Noise monitoring, 61–62
Noise Reduction Rating (NRR), 62
Non-emergency care, 88

Occupational physician, 37, 84
Orientation, 89, 92
Overprotection, 94

Permeation, 152, 153, 157
Permissible exposure limit, 85, 86, 139
Personal protective equipment (PPE), 64, 80, 94, 107–148
Physical hazards, 77–78
Process hazard analysis, 52
Process safety, 18, 227–248
Program and course evaluations, 101
Project manager (PM), 32, 33, 41, 55, 69, 108, 223
Protection factor, 141
Purchase order, 220

Radioactive materials, 1
Radiological hazards, 59–60
Refusal (acknowledgement), 75
Remediation, 6
Respiratory protection, 34, 132–139
Review, 38
Rinsing, 155

Safety alert, 39, 40
Safety culture, defined, 3

Safety meetings, 34
Sanitation, 161–162
Scaffolds, 58
Security, 37, 38
Shower area, 161
Site control/work zones, 81
Site health and safety officer (SSHO), 33–36
Site inspection, 90
Site manager, 33, 41
Site supervisor, 108
Solidification, 155
Sterilization, 156
Subcontractors. *See* Contractors
Supervised field experience, 98
Support personnel, 23
Support zone, 65
Surfactants, 155

Training, 7–8, 42–49, 82, 90, 96–107
Training certification, 99

Upgrading/downgrading levels of protection, 34, 120–123

Visitors, 36

Warning properties, 141
Waste minimization, 64, 162–163
What if, 50
Wipe sampling, 156
Work plan, 39
Worker comfort areas, 66